RF Analog Impairments Modeling for Communication Systems Simulation

T0327496

RF Analog Impairments Modeling for Communication Systems Simulation
Application to OFDM-based Transceivers

Lydi Smaini

Marvell Switzerland

A John Wiley & Sons, Ltd., Publication

This edition first published 2012
© 2012, John Wiley & Sons Ltd

Registered office
John Wiley & Sons Ltd, The Atrium, Southern Gate, Chichester, West Sussex, PO19 8SQ, United Kingdom

For details of our global editorial offices, for customer services and for information about how to apply for permission to reuse the copyright material in this book please see our website at www.wiley.com.

Library of Congress Cataloging-in-Publication Data

Smaini, Lydi, 1974-
   RF analog impairments modeling for communication systems simulation : application to OFDM-based transceivers / Lydi Smaini.
        p. cm.
     ISBN 978-1-119-99907-2 (hardback)
   1. Radio–Transmitter-receivers–Simulation methods. 2. Electromagnetic interference. 3. Signal integrity (Electronics) 4. Telecommunication systems–Simulation methods. 5. Orthogonal frequency division multiplexing. I. Title.
     TK7867.2.S63 2012
     621.382–dc23

                                                            2012016441

A catalogue record for this book is available from the British Library.

Print ISBN: 9781119999072

Typeset in 10/12.5 Palatino by Laserwords Private Limited, Chennai, India

To my parents, Zina and Rabah.
To my sisters, Nadine and Assia.
To my brothers, Malik and Dalil.
To my childhood friend, Djamel.

Contents

Preface

Modern digital communications transceivers can be decomposed into two main parts: the radio frequency (RF) analog front-end, which transmits and receives the analog signal, and the digital baseband, which is responsible for the digital signal processing (DSP) and data demodulation. The "virtual" frontier is delimited by the digital to analog conversion in transmission, and by the analog to digital conversion in reception. The digital baseband is commonly studied and simulated by communication system and DSP engineers based on standard requirements, which specify the modulation type and the system performance in terms of bit or packet error rate. On the other hand, the RF analog front-end specifications are often derived by the RF analog engineers themselves using, for example, Excel spreadsheets for calculating signal-to-noise ratio (SNR) or error vector magnitude (EVM) from basic and classical formulas based on dual- or single-tone tests historically coming from laboratory measurements. Actually, these RF analog analysis methods do not take into account the signal spectral properties and the transceiver bandwidth; consequently, there is a gap between the RF analog front-end specifications and the digital baseband simulations which often introduces misunderstanding during the transceiver design.

Nowadays, with the growing complexity of personal mobile communication systems demanding higher data-rates and high levels of integration using low-cost complementary metal oxide semiconductor (CMOS) technology, overall system performance has become much more sensitive to RF analog font-end impairments. Consequently, communication system and DSP engineers have to understand and to include these RF analog imperfections in their simulation benches in order to measure their impact on the system performance. In addition, in deep-submicrometer CMOS technology (nanometer) the digital part of the transceiver naturally shrinks with the process ratio whereas the RF analog part remains fairly constant (only 10 % size reduction in good cases). As a result, in terms of die area and thus cost reduction, the analog part remains the major bottleneck of CMOS transceiver integration and generally requires a non-negligible redesign effort if one wants to reduce its area and power consumption. To surmount this integration issue, a new trend is to design suboptimal RF analog front-ends, called "dirty RF" in recent literature, and to compensate their impairments with DSP. Designing such integrated transceivers requires a thorough understanding of the whole transceiver chain, including RF analog engineering and DSP.

The aim of this book is to provide the reader with theoretical and practical RF analog system modeling knowledge and examples directly applicable to advanced transceiver studies and simulations. Furthermore, we endeavor to make a bridge between RF analog designers and communication system/DSP engineers who often use different tools and vocabulary even when specifying the same thing. We theoretically describe the impact of the RF analog imperfections on orthogonal frequency division multiplexing (OFDM) modulation, which has widely recognized advantages and is utilized in latest generation communication systems. To illustrate the theory we present simulation results comparing the impact of the transceiver imperfections on two well-known deployed standards, WiFi (802.11a/g) and mobile WiMAX (802.16e).

The organization of the book is as follows:

- In Chapter 1 we introduce the challenges of the integration of communication systems on-chip, especially those using CMOS technology. We will also give an overview of the major RF analog front-end architectures, the main principles of OFDM modulation which is now deployed in 4G mobile communications, and finally an introduction to RF analog system performance metrics and baseband simulation.
- Chapter 2 deals with the principal RF analog impairments encountered in communication transceivers. We describe them mathematically in order to derive models which can be incorporated into any system simulation, and also to study their theoretical impact on the system performance with a special focus on OFDM modulation.
- In Chapter 3, system simulation results based on WiFi and mobile WiMAX OFDM transceivers are presented. All the RF analog impairments described in Chapter 2 are modeled and simulated in order to study their impact individually on the system performance, both in transmission and reception.
- Finally, digital estimation and compensation of the RF analog impairments is overviewed in Chapter 4: carrier and sampling frequency offsets in OFDM reception, quadrature imbalance, as well digital pre-distortion in transmission are addressed.

Acknowledgments

First, I would like to thank Frederic Declercq, Analog IC design manager at Marvell Switzerland, and Kevin Koehler, Staff DSP engineer at Marvell Switzerland, for having accepted to review the whole manuscript. Their valuable comments on both content and form, and our technical discussions allowed me to improve the book from its original version to the final delivery. Thanks also to Nils Rinaldi, Project manager at EPFL, for his comments on Chapter 2.

I am grateful to Patrick Clement, Director of Marvell Switzerland, for his trust and encouragement to write this book.

I wish to thank Peter Mitchell, Publisher at John Wiley & Sons responsible for electrical and electronics engineering books, who proposed this book project to me. I also thank Liz Wingett, Project Editor at John Wiley & Sons, for her support and advice during the writing process.

Finally, I would like to thank my fiancée Laurence for her understanding, patience, and support.

About the Author

Dr Lydi Smaini was born in Tizi-Ouzou, Algeria, on 27 July 1974. He received his M.S. and Ph.D. degrees in electronics from the University of South Toulon-Var, France, in 1998 and 2001, respectively, specializing in radio propagation, telecommunications, and remote sensing. His thesis work focused on pulse compression techniques and signal processing for atmospheric radars.

After graduation he worked for one year as an R&D electronics engineer for ALTEN, Marseille, France, where he developed a frequency agile radar transponder beacon (S and X bands) for navigation aid.

From 2002 to 2006 he was with STMicroelectronics in the RF System and Architecture Group for wireless communications, Geneva, Switzerland, where he worked on ultra wide-band impulse radio, 3G cellular phones, and advanced radio architectures for orthogonal frequency division multiple access (OFDMA) technology.

Dr Smaini joined Marvell Switzerland in July 2006, Etoy, Switzerland, where he is currently leading the RF System and DSP group working on deep sub-micrometer complementary metal oxide semiconductor (CMOS) telecommunication transceivers.

1

Introduction to Communication System-on-Chip, RF Analog Front-End, OFDM Modulation, and Performance Metrics

1.1 Communication System-on-Chip

1.1.1 Introduction

Radio frequency (RF) communication systems use RFs to transmit and receive information such as voice and music with FM, or video with TV, and so on (Steele, 1995; Rappaport, 1996; Haykin, 2001). From a general point of view RF communication is simply composed of an RF transmitter sending the information and an RF receiver recovering the information (Figure 1.1). Below are basic definitions of the vocabulary commonly used in communication systems:

- **Signal:** Information (data, image, music, voice, . . .) we want to transmit and receive.
- **Carrier frequency**: RF sinusoidal waveform, called a carrier because it is used to "carry" the signal from the transmitter to the receiver.
- **MODulation**: Modifying the carrier waveform in order to convey the information (signal) in transmission.

RF Analog Impairments Modeling for Communication Systems Simulation:
Application to OFDM-based Transceivers, First Edition. Lydi Smaini.
© 2012 John Wiley & Sons, Ltd. Published 2012 by John Wiley & Sons, Ltd.

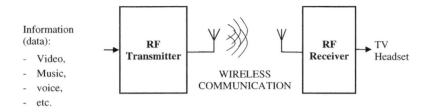

Figure 1.1   Basic view of an RF communication system

- **DEModulation**: Extracting the signal (i.e., the information) from the carrier frequency in reception.
- **Antenna**: Device which transforms the electrical signal into electromagnetic waves for radiation and vice versa.
- **Channel bandwidth**: Span of frequencies used for the communication.
- **MODEM** = MODulator + DEModulator.
- **TRANSCEIVER** = TRANSmitter + reCEIVER.

In the last decades telecommunications have migrated toward digital technology (Proakis, 1995) as a result of the evolution of advanced digital signal processing (DSP) techniques which can now be deployed at low-cost in mobile devices. Nowadays a mobile phone is not only used for traditional voice calls but as a multimedia platform for surfing the Internet, listening to music, data transfers, localization (global positioning system (GPS)), and so on: many applications which require the implementation of different technologies and communication standards (WiFi, Bluetooth, GSM/3G/4G Long Term Evolution (LTE), GPS, near-field communication (NFC), etc.) on the same platform. Since the phone's form factor and battery life are limited, state-of-the-art integrated circuit (IC) design and system-on-chip (SoC) implementations have become necessities for providing cost-effective solutions to the market.

Modern digital communications transceivers (Figure 1.2) are generally composed of a Medium Access Control (MAC) layer managing the access to the medium between different users in a network and the quality of service seen by each, and a PHY (Physical

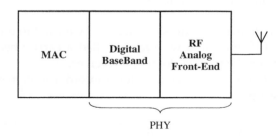

PHY

Figure 1.2   Basic partitioning of a digital communication transceiver

Layer) which is responsible for the transfer of information across the medium (wireless channel, cable, optical fiber, etc.). The PHY can be decomposed into two blocks:

- The digital baseband (DBB) which is located between the MAC and the analog front-end (AFE). The baseband transmission path encodes the bits provided by the MAC, generates the data symbols to be sent across the medium, and finally performs the digital modulation. The reception path demodulates the data and provides a decoded bit stream to the MAC. Generally, the transmission requirements are well specified by the standards (channel coding, modulation, etc.), whereas the algorithms used in reception (channel estimation/equalization, synchronization, etc.) can vary from one implementation to another.
- The RF AFE is connected to the DBB. The RF transmit path converts the DBB signal to analog and frequency up-converts to RF. The receiver frequency down-converts the RF signal to baseband, filters out any interferers, and finally converts the signal to DBB.

1.1.2 CMOS Technology

As complementary metal oxide semiconductor (CMOS) technology presents remarkable shrinking properties and cost attractiveness, it has become the unavoidable choice for semiconductors implementing SoC and for low-cost combo-chips integrating several systems on the same die (Abidi, 2000; Brandolini *et al.*, 2005). Although CMOS was initially dedicated to digital design, today RF AFEs are embedded using this technology as well in order to improve the integration efficiency and thus lower the platform cost (Lee, 1998; Razavi, 1998a,b; Iwai, 2000). Nevertheless, CMOS is not well-optimized for RF analog design due to the low ohmic substrate limiting the analog/digital isolation, the low-voltage supply limiting the dynamic range/linearity, and the poor quality factor of the passive components. Furthermore, in deep-submicrometer CMOS technology (nanometer), whereas the digital part of the chip naturally shrinks with the process ratio, the RF analog part scales poorly (Figure 1.3), at around 10% per process node, and generally requires a redesign in order to be able to reduce its area and power consumption. Consequently, for SoC integration the RF AFE remains the major bottleneck in reducing the CMOS transceiver size, therefore requiring more work.

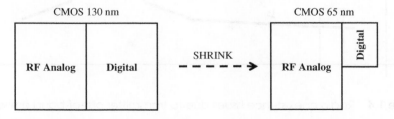

Figure 1.3 SoC shrink limitation due to the RF analog part of the chip

1.1.3 Coexistence Issues

Due to the integration constraints imposed by multi-communication applications, several communication systems often have to coexist on the same platform (such as mobile phone), and in the worst case even on the same chip. Even if the radios do not operate in the same band, any RF transmitter generates broadband out-of-band emissions which can degrade the sensitivity of neighboring receiver bands, as illustrated in Figure 1.4.

If the systems are located on the same platform but not on the same chip, a coupling between antennas, or between chips at the pin level, can occur, as depicted in Figure 1.5. Board design and layout, as well as the distance between the antennas and their orientation, have to be carefully taken into account for limiting the coupling factor between the two systems.

The most difficult case concerns the recent combo-chips, in which the different communication systems are embedded on the same die (Figure 1.6), especially if the power amplifiers (PAs) are also integrated. In addition to the external coupling, on-chip leakage and coupling can pose particular problems because the RF filtering is not present at this level. As with the board design, chip layout and position of the blocks are fundamental design considerations.

Because modern receiver sensitivities are generally specified to be very low for guaranteeing good reception even in weak signal conditions and to relax the transmit power requirements, transmitter out-of-band emissions can rapidly become a real bottleneck if they increase the overall noise floor of the multi-communications system. For example, let us suppose a victim receiver having a bandwidth of $1\,\mathrm{MHz}$ and a noise figure of $5\,\mathrm{dB}$; in this case its input-referred noise floor is $-109\,\mathrm{dBm}/\mathrm{MHz}$ ($kTBF$ at room temperature: $-174 + 10\log_{10}(1e6) + 5$). By assuming a coupling factor between

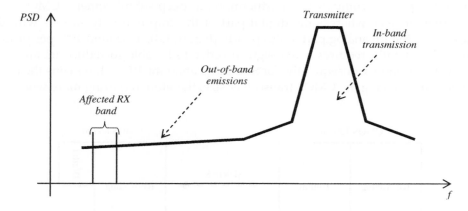

Figure 1.4 Radio coexistence issues due to transmitter out-of-band emissions

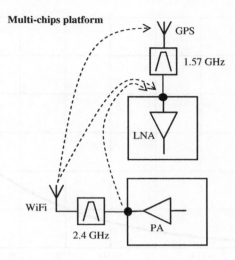

Figure 1.5 Multi-chips coupling through the antennas

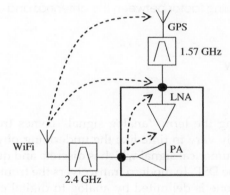

Figure 1.6 Antennas and systems on-chip coupling

the antennas, we can estimate its sensitivity degradation as a function of the transmitter out-of-band emissions level seen in the receiver band. Figure 1.7 is a plot of the victim receiver sensitivity degradation as function of the transmitter out-of-band emissions assuming a coupling factor of −10 dB between the two antennas. The degradation is negligible, that is, smaller than 0.1 dB, if the transmitter out-of-band emissions are lower than −115 dBm/MHz in the receiver bandwidth. The emission specification can be relaxed if we tolerate a higher degradation; for example, the transmitter out-of-band emissions can reach −105 dBm/MHz for an allowed degradation of 1 dB.

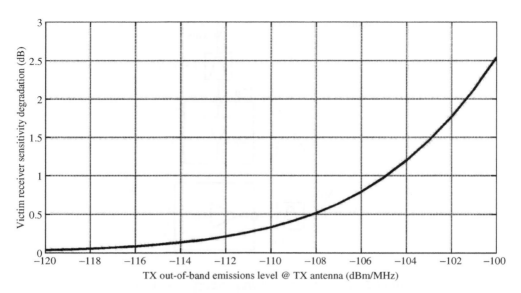

Figure 1.7   RX sensitivity degradation as a function of the TX out-of-band emissions level by assuming −10 dB coupling factor between the antennas and −109 dBm/MHz receiver noise floor

1.2 RF AFE Overview

1.2.1 Introduction

In electronics engineering the term "analog signal" comes from the ANALOGY of the signal to continuously vary in time like the underlying physical phenomenon, as opposed to the digital/numerical signal which is discrete and quantized with a certain resolution imposed by the DSP. In modern transceivers the frontier between the analog domain and the digital one is delimited by analog to digital converters (ADCs) and digital to analog converters (DACs) in reception and transmission, respectively, as depicted in Figure 1.8. The RF AFE is composed of two paths, one for receiving the RF signal and the other one for transmitting the baseband signal.

The primary function of an RF analog receiver is to amplify and to frequency down-convert the desired signal from RFs to baseband with minimum degradation. Its main requirements are:

- The frequency band, imposed by the regulation bodies, specifying the local oscillator (LO) frequency range.
- The signal bandwidth specifying the analog baseband filtering.
- The minimum sensitivity specifying the noise figure imposed by the minimum signal-to-noise ratio (SNR) required for good data demodulation.

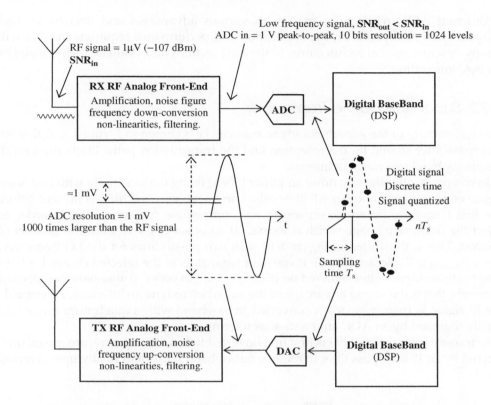

Figure 1.8 RF analog and DBB partitioning in modern transceivers

- The signal dynamic range specifying the automatic gain control (AGC) design and the ADC resolution.
- The adjacent channel selectivity (ACS) and the blockers/interferer rejection specifying the linearity, the analog baseband filtering, and the LO phase noise profile.

The principal function of an RF analog transmitter is to frequency up-convert the baseband signal to RF and to amplify it to the required transmission power. Its major requirements are:

- The frequency band specifying the LO frequency range.
- The maximum transmission power specifying the PA.
- The error vector magnitude (EVM); that is, modulation accuracy, specifying the noise budget including thermal noise, phase noise, DAC resolution, and linearity.
- The adjacent channel leakage ratio (ACLR) specifying the spectrum emission mask.
- The spurious emission mask specifying the out-of-band unwanted emissions.

Different AFE architectures exist with various advantages and drawbacks, and proper selection depends on the application and performance requirements. We will briefly describe several architectures in the next sections and evaluate their suitability for SoC integration.

1.2.2 Superheterodyne Transceiver

The architecture of the superheterodyne transceiver is shown in Figure 1.9. A first RF bandpass filter shared by the reception and the transmission paths limits the overall frequency band to a range of interest.

In reception, after the low-noise amplifier (LNA) fixing the sensitivity with low noise figure and high gain, another RF filter called an image rejection filter is present before the first frequency down-conversion to an intermediate frequency (IF), in order to reject the frequency image which is located at an offset of $2 \times$ IF from the channel of interest. This is illustrated in Figure 1.10 with two possibilities for the LO frequency: $f_{LO} = f_{channel} \pm$ IF. Because the IF is constant regardless of the selected channel, a high selectivity bandpass filter centered on IF can be used in order to attenuate the adjacent channels; this is the strong advantage of the superheterodyne architecture. Afterwards the IF signal is frequency down-converted to baseband with a quadrature mixer and finally digitized by an ADC after anti-alias filtering.

In transmission the channel is first up-converted to IF with a quadrature mixer, then filtered by an IF bandpass filter to remove out-of-band noise, and finally up-converted

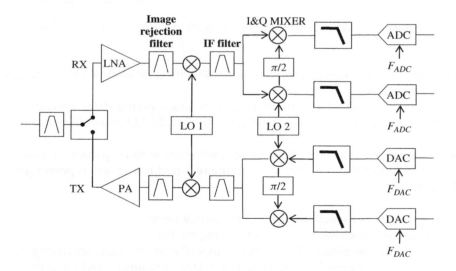

Figure 1.9 Superheterodyne transceiver

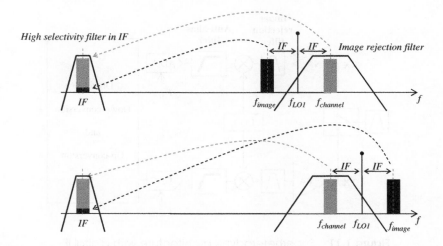

Figure 1.10 Frequency down-conversion in IF and frequency image issue

to the desired RF channel frequency with the second LO. Before the PA an image rejection filter is necessary in order to attenuate the image before the antenna.

Although the superheterodyne architecture is very well known and presents interesting advantages like high selectivity and limited LO leakage/pulling, it is not appropriate for on-chip integration because of its complexity (two LOs) and the number of components which are prohibitive in low-cost solutions. For example, the two bandpass filters (image rejection and IF) are difficult to integrate on-chip because they require high quality factor components not easily obtained in CMOS technology.

An alternative for limiting the use of discrete analog components is to digitize the real bandpass channel at IF and to perform the quadrature frequency down-conversion and up-conversion to baseband in digital. The superheterodyne architecture with digital IF is presented in Figure 1.11. We can note that this transceiver requires only one DAC and one ADC. The main motivation of this architecture is to avoid the analog IF filter which is very difficult to integrate on-chip. The complexity of this architecture is now dominated by the ADC performance because it has to handle signal levels which include the adjacent channels and to sample the signal at higher frequency (IF), meaning high dynamic-range requirements and more sensitivity to clock jitter. In addition, this architecture still requires an image rejection filter between the LNA and the mixer which is difficult to integrate on-chip at low cost. An option is to increase the IF in order to relax the image rejection filter order, but in this case the DAC and ADC specifications become even more stringent. Consequently, like the classical superheterodyne, this digital IF architecture is not really suitable for low-cost solution integration in CMOS.

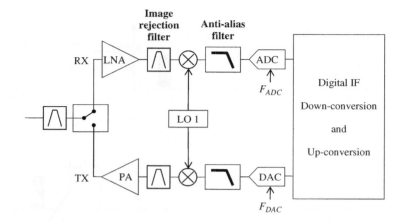

Figure 1.11   Superheterodyne architecture with digital IF

1.2.3 Homodyne Transceiver

A block diagram of the homodyne transceiver (Tucker, 1954) is depicted in Figure 1.12. It is also known as a zero-IF transceiver. The major difference with the superheterodyne transceiver is that the channel, that is, the desired signal, is directly frequency down-converted to baseband in reception, or frequency up-converted to RF in transmission, without using an IF, as shown in Figure 1.13. It is for that reason the homodyne

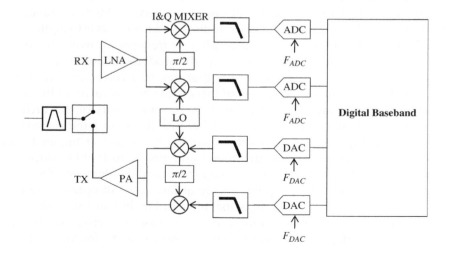

Figure 1.12   Homodyne or zero-IF architecture

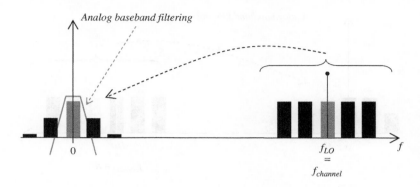

Analog baseband filtering

f_{LO}
=
$f_{channel}$

f

0

Figure 1.13   Frequency down-conversion to DC in homodyne or zero-IF architecture

architecture is also called a direct conversion transceiver, because the LO is directly tuned to the desired channel frequency.

Because there is no IF, the RF image rejection filters and the IF bandpass filters are not necessary anymore, thus simplifying the architecture. Consequently, zero-IF architecture is a good candidate for SoC integration (Abidi, 1995; Razavi, 1997); however, different drawbacks exist in both transmission and reception, especially due to LO leakage. Because in reception the baseband signal is centered on DC, it is very sensitive to DC offset coming from the LO self-mixing and baseband active devices. The DC offset can be considerable compared with the desired signal and can therefore severely affect the dynamic range requirements of the baseband path. Regarding on-chip integration, we will see in the next chapter that flicker noise, especially in CMOS, can also limit the performance of narrow-band receivers.

In transmission, the LO leakage into the channel and the LO pulling due to the PA output can be serious performance bottlenecks.

1.2.4 Low-IF Transceiver

As we have previously seen, the zero-IF architecture is a good option for SoC integration but suffers from DC offset issues in reception and LO leakage sensitivity in transmission because the LO and the signal are at the same frequency. One solution is to use a low-IF architecture in which the LO is slightly shifted compared with the channel (Crols and Steyaert, 1998). Low-IF architecture is similar to the zero-IF one depicted in Figure 1.12. The main difference is the fact that the desired channel is offset from DC, requiring wider analog baseband filters than the channel bandwidth, as illustrated in Figure 1.14 for the receiver case. It is interesting to note in this figure that the neighboring channels are included in the baseband band power, which will impact the ADC dynamic range and the AGC. The channel selection is done in digital, where the DC offset is also removed.

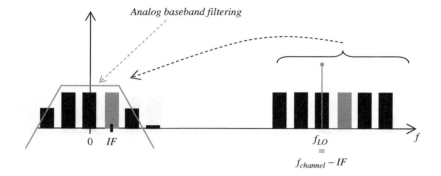

Figure 1.14   Frequency down-conversion to IF in low-IF architecture

1.2.5 Analog Baseband Filter Order versus ADC Dynamic Range

One of the chronic questions for die area and power reduction in the design of CMOS AFEs is the trade-off in reception between the analog baseband filtering order and the ADC dynamic range (Figure 1.15).

By using a high-order analog baseband filter, the out-of-band blockers and interferers are attenuated before the ADC, as illustrated in the Figure 1.16a. The major part of the channel selectivity is achieved in analog and the ADC dynamic range can be optimized for the required signal-to-noise plus distortion ratio (SNDR). In this case the design constraints are especially put on the analog baseband filtering.

In modern transceiver architectures, high dynamic range ADCs allow the use of lower filter orders, in which case the power of the blockers contributes more at the ADC input. As we can see in the Figure 1.16b, the signal is squeezed because the receiver gain must be adjusted based on the blockers strength. In addition we have to be careful about the receiver linearity and noise figure in order to always guarantee the required SNDR. In these new transceiver architectures, the selectivity is especially provided by the DSP.

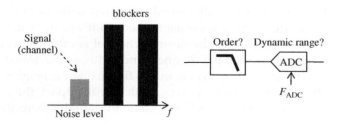

Figure 1.15   Trade-off between analog baseband filter order and ADC dynamic range

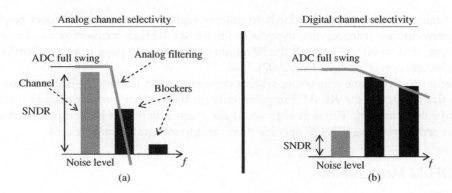

Figure 1.16 Analog baseband filter order impact on the receiver ADC dynamic range

Another point to take into consideration is the sampling of the analog baseband signal as part of the digitization by the ADC. Indeed, aliasing occurs if spectral components are present above half of the sampling frequency, as demonstrated by the Nyquist–Shannon sampling theorem.

Because the signal of interest is located at baseband, the components around nF_s will be aliased within the channel as depicted in Figure 1.17. Consequently, the analog baseband filtering must not only be dimensioned as a function of the ADC dynamic range as discussed earlier, but also to limit the aliasing of blockers and noise located at multiples of the ADC clock frequency into the DBB.

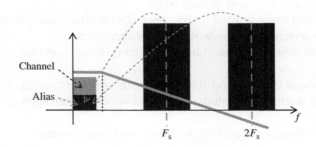

Figure 1.17 Blockers aliasing before the ADC vs. analog baseband filtering order

1.2.6 Digital Compensation of RF Analog Front-End Imperfections

As we have seen in Section 1.1.2, the integration of the RF AFE on-chip is a key point for providing low-cost solutions by reducing the number of chips and saving power in modern mobile handheld devices. Today, CMOS technology is the best candidate for low-cost SoC development but is more adapted to digital design and not really optimized for analog design. In addition, the trend of new digital telecommunication standards to deliver higher data-rates to the user requires high-performance transceivers, especially for the RF AFE in terms of noise, linearity, matching, and so on. As a result the RF analog impairments pose a serious bottleneck to the integration (Fettweis *et al.*, 2007).

In order to overcome this issue, system designers need to deeply study and understand the impact of the RF AFE impairments on the system performance in order to quantify their impact. The aim is to see if the most severe can be compensated with DSP in order relax the RF AFE specifications and to integrate it at low cost.

1.3 OFDM Modulation

1.3.1 OFDM as a Multicarrier Modulation

Orthogonal frequency division multiplexing (OFDM) is a multicarrier modulation increasingly used in communication systems (WiFi, WiMAX (Worldwide Interoperability for Microwave Access), 4G/LTE, power line communication (PLC)) because it presents several advantages compared with single carrier modulation and classical frequency division multiplexing (Bingham, 1990; Van Nee and Prasad, 2000; Prasad, 2004; Li and Stuber, 2006; Armstrong, 2007):

- Efficient use of the spectrum because subcarrier orthogonality allows overlap.
- Less sensitive to channel fading (multipath).
- Channel estimation and equalization in the frequency domain carries low complexity.
- In frequency-selective fading possibility to avoid the affected subcarriers or to adapt their modulation as a function of their SNR.
- Possibility to avoid inter-symbol interference (ISI) with a cyclic prefix.
- Narrow band interferers will only affect few subcarriers.
- Coexistence with other systems: subcarriers can be turned on/off.

However, the use of OFDM modulation presents some drawbacks which have to be taken into account during the system design and specification, such as:

- Very sensitive to frequency and phase offsets and timing error:
  - Break the orthogonality between subcarriers.
- OFDM temporal signal has a high peak to average power ratio (PAPR):
  - Poor efficiency of the PAs.
  - Signal clipping and distortion degrades the SNR and generates out-of-band emissions.

Figure 1.18 schematically describes the principle of OFDM modulation and demodulation, in baseband with subcarriers around DC, using an inverse Fourier transform (IFT) in transmission and a Fourier transform (FT) in reception.

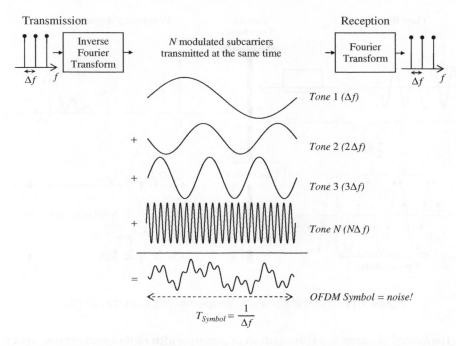

Figure 1.18   Basic principle of OFDM transmission and reception (baseband illustration)

In transmission, the data are modulated (or mapped) onto N subcarriers at the input to the IFT which generates an OFDM baseband symbol in the time domain whose duration is inversely proportional to the subcarrier spacing. Because the OFDM symbol is composed of a large number of modulated subcarriers, it looks like a Gaussian noise process in the temporal domain under the central limit theorem, thus explaining its high PAPR.

In reception, after time synchronization to find the beginning of the OFDM symbols, an FT is applied for recovering the subcarriers and then demodulating the data. In reality the subcarriers received are distorted in amplitude and phase by a real transmission channel; as a result channel estimation and equalization are needed for compensating the FT output result before data demodulation.

1.3.2 Fourier Transform and Orthogonal Subcarriers

As we have seen in the previous section, the OFDM modulation is performed using the FT mathematical operator (Weinstein and Ebert, 1971). Figure 1.19 is a review of basic FT properties which are necessary to bear in mind for understanding the OFDM modulation principle.

Because OFDM subcarriers are generated during a limited duration T, or equivalently infinite sine waves multiplied by a rectangular window in the time domain, the result

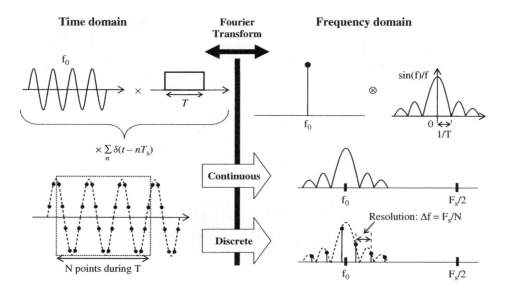

Figure 1.19   Fourier transform properties important in OFDM

in the frequency domain is a convolution of a zero-width delta function (the spectrum of the infinite-time subcarriers) by the FT of the rectangular window, which is merely a sinc function. The continuous FT of the OFDM symbol $s(t)$ is given by

$$s(t) = w(t) \times \sum_k s_k \, e^{j2\pi k \Delta ft} \xleftrightarrow{\text{FT}} S(f) = \int_t s(t) \, e^{-j2\pi ft} \, dt \tag{1.1}$$

which can be decomposed as a convolution product:

$$S(f) = \int_t w(t) \, e^{-j2\pi ft} \, dt \otimes \int_t \sum_k s_k \, e^{j2\pi k \Delta ft} \, e^{-j2\pi ft} \, dt \tag{1.2}$$

giving

$$S(f) = W(f) \otimes \sum_k s_k \delta(f - k\Delta f) = \sum_k s_k W(f - k\Delta f) \tag{1.3}$$

in which $w(t)$ and $W(f)$ are the window responses in the time and frequency domains, respectively, $\delta()$ is the Dirac function, $\otimes$ is the convolution operator, s_k is the complex symbol modulating subcarrier k, and Δf is the subcarrier spacing.

If $w(t)$ is a rectangular window having a duration T, defined between $-T/2$ and $T/2$, we can rewrite Equation 1.3 as

$$S(f) = T \sum_k s_k \frac{\sin[\pi (f - k\Delta f)T]}{\pi (f - k\Delta f)T} \tag{1.4}$$

Equation 1.4 shows that a sinc function, introduced by the FT of the rectangular window, is centered on each modulated subcarrier. The final spectrum is merely the sum of all these shifted and scaled sinc functions.

Because the OFDM transceiver uses a discrete Fourier transform (DFT) applied on N samples, defined by the number of subcarriers, Equation 1.1 becomes

$$s(nT_s) = s(t) \times \sum_{n=0}^{N-1} \delta(t - nT_s) \xleftrightarrow{\text{DFT}} S\left(\frac{m}{NT_s}\right) = \frac{1}{N} \sum_{n=0}^{N-1} \sum_{k=-N/2}^{N/2-1} s_k \, e^{j2\pi k\Delta f n T_s} \, e^{-j\frac{2\pi}{N}mn} \quad (1.5)$$

in which T_s is the sampling period and m is the DFT frequency bin index.

For more clarity in the following development, we rewrite Equation 1.5:

$$S\left(\frac{m}{NT_s}\right) = \frac{1}{N} \sum_{k=-N/2}^{N/2-1} s_k \sum_{n=0}^{N-1} e^{-j\frac{2\pi}{N}(m-kN\Delta fT_s)n} \quad (1.6)$$

By using the result of the geometric series

$$\sum_{n=0}^{N-1} z^{-n} = \frac{1 - z^{-N}}{1 - z^{-1}} \quad (1.7)$$

we can develop Equation 1.6

$$S\left(\frac{m}{NT_s}\right) = \frac{1}{N} \sum_{k=-N/2}^{N/2-1} s_k \frac{1 - e^{-j2\pi(m-kN\Delta fT_s)}}{1 - e^{-j\frac{2\pi}{N}(m-kN\Delta fT_s)}}$$

$$= \sum_{k=-N/2}^{N/2-1} s_k \frac{\sin\left[\pi\left(m - kN\Delta fT_s\right)\right]}{N \sin\left[\frac{\pi}{N}\left(m - kN\Delta fT_s\right)\right]} \, e^{-j\pi\left(\frac{N-1}{N}\right)(m-kN\Delta fT_s)} \quad (1.8)$$

In order to ease the study of Equation 1.8, the sum can be decomposed into two components to analyze separately ($m = k$ and $m \neq k$):

$$S\left(\frac{m}{NT_s}\right) = S_{m=k}\left(\frac{m}{NT_s}\right) + S_{m \neq k}\left(\frac{m}{NT_s}\right) \quad (1.9)$$

$m = k$: Considered subcarrier

$$S_{m=k}\left(\frac{k}{NT_s}\right) = s_k \frac{\sin\left[k\pi(1 - N\Delta fT_s)\right]}{N \sin\left[\frac{k\pi}{N}(1 - N\Delta fT_s)\right]} e^{-jk\pi\left(\frac{N-1}{N}\right)(1-N\Delta fT_s)} \quad (1.10)$$

Equation 1.10 shows that all the original symbols s_k will be distorted in amplitude and in phase depending on the ratio of the subcarrier frequency Δf to the sampling frequency $F_s = 1/Ts$.

$m \neq k$: Inter-carrier interference (ICI)

$$S_{m \neq k}\left(\frac{m}{NT_s}\right) = \sum_{\substack{k=-N/2 \\ m \neq k}}^{N/2-1} s_k \frac{\sin\left[\pi\left(m - kN\Delta fT_s\right)\right]}{N \sin\left[\frac{\pi}{N}(m - kN\Delta fT_s)\right]} \, e^{-j\pi\left(\frac{N-1}{N}\right)(m - kN\Delta fT_s)} \tag{1.11}$$

Equation 1.11 shows that all the subcarriers will be distorted by the $(N-1)$ others, which is referred to as inter-carrier interference (ICI), and, when present, limits the orthogonality of the OFDM modulation.

In order to obtain orthogonal subcarriers at the DFT output (Equation 1.6) there must be neither subcarrier distortion nor ICI present, which can be mathematically expressed as

$$\begin{aligned} S_{m=k}\left(\frac{k}{NT_s}\right) &= s_k \\ S_{m \neq k}\left(\frac{m}{NT_s}\right) &= 0 \end{aligned} \tag{1.12}$$

The condition for no ICI is

$$\pi\left(m - kN\Delta fT_s\right) = n\pi \tag{1.13}$$

for $m \neq k$ and $n \neq 0$, which is satisfied for

$$N\Delta fT_s = 1 \Rightarrow \Delta f = \frac{1}{NT_s} \tag{1.14}$$

Equation 1.14 indicates that the subcarriers are orthogonal if their frequencies are integer multiples of the DFT frequency resolution. In practice, this means that when the DFT computes the considered subcarrier k, that is, $m = k$, the nulls of the sinc function are perfectly aligned with the other subcarriers, producing no ICI.

The previous equations establishing the condition to obtain subcarrier orthogonality are visually summarized in Figure 1.20. We can clearly see the DFT weighting function and the ICI effect if the subcarriers are not multiples of the DFT frequency resolution.

We will see in the subsequent chapters that the RF analog impairments such as sampling and carrier frequency offsets will violate this condition.

1.3.3 Channel Estimation and Equalization in Frequency Domain

An important factor affecting the communication systems performance is the propagation channel between the transmitter and the receiver. In wireless communications the signal is received several times with different delays and strengths due to the multipath reflection, and can also be frequency shifted because of the Doppler effect. From a

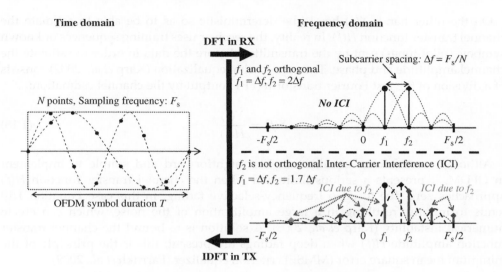

Figure 1.20 Subcarrier orthogonality and ICI in OFDM

signal processing point of view, the received signal is the result of the convolution of the transmitted signal by the channel impulse response:

$$y(t) = x(t) \otimes h(t) = \int_\tau x(\tau)h(t-\tau)\,\mathrm{d}\tau \tag{1.15}$$

where $y(t)$ and $x(t)$ are the received and the transmitted signals, respectively, $h(t)$ is the channel impulse response, and $\otimes$ is the convolution operator.

Consequently, before the data demodulation the receiver has to compensate the signal distortion due to the channel; it is performed through channel estimation and equalization. Whereas in classical single-carrier systems the channel estimation and equalization have to deal with the convolution of the signal by the channel in time domain, OFDM receivers transform this convolution into multiplication with the FT (Van de Beek *et al.*, 1995; Morelli and Pun, 2007):

$$y(t) = \int_\tau x(\tau)h(t-\tau)\,\mathrm{d}\tau \xleftarrow{\text{Fourier transform}} Y(f) = X(f) \times H(f) \tag{1.16}$$

in which $Y(f)$, $X(f)$, and $H(f)$ are the spectra of $y(t)$, $x(t)$, and $h(t)$, respectively. $H(f)$ is also called a channel transfer function.

We clearly see in Equation 1.16 the advantage of the channel estimation in frequency domain because the temporal deconvolution to extract the channel is transformed into a simple division:

$$H(f) = \frac{Y(f)}{X(f)} \tag{1.17}$$

On the other hand, $X(f)$ has to be deterministic so as to be able to estimate the channel transfer function $H(f)$. In reality, the receiver uses training sequences or known symbols (pilot-tones) sent by the transmitters before the data in order to estimate the channel amplitude and phase. The zero-forcing equalization (Karp *et al.*, 2002) consists of a division of the fast Fourier transform (FFT) output by the channel estimation:

$$X(f) = \frac{Y(f)}{H(f)} \tag{1.18}$$

Although zero-forcing equalization is straightforward and simple to implement in OFDM, it presents a serious limitation when the channel transfer function $H(f)$ approaches zero due to deep frequency-selective fading. In this case Equation 1.18 tends to infinity, resulting in a large amplification of the noise, which can create numerical instability (Karp *et al.*, 2003). A solution is to bound the channel transfer function amplitude $H(f)$ when deep fadings are present; this is the principle of the minimum mean square error (MMSE) criterion equalizer (Farrukh *et al.*, 2009).

1.3.4 Pilot-Tones

In order to aid the receiver in estimating the propagation channel and to correct transceiver impairments such as carrier frequency and sampling clock offsets between transmitter and receiver, deterministic subcarriers are sent by the transmitter (Figure 1.21). These carriers are called pilots (Hoeher *et al.*, 1997; Tufvesson and Maseng, 1997). Knowing their properties, that is, their frequencies and their phase and amplitude, the receiver can exploit them in order to extract the phase and amplitude of the channel to be used by the equalizer to compensate the output of the FFT. Because the demodulated data quality strongly depends on the channel estimation precision, the demodulation of the pilots has to be robust. Consequently, they are generally boosted in power a few decibels above the subcarriers used for data.

Because the pilots are much less than the total number of subcarriers, the channel estimation for all the subcarriers is performed using interpolation methods.

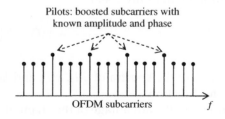

Figure 1.21   Pilot-tones used in reception for channel estimation

It is also possible to track the channel in time and frequency by moving the pilot positions between OFDM symbols.

1.3.5 Guard Interval

In communication systems the multipath reflections due to signal propagation create channel delay spread which can severely affect receiver demodulation performance. This is known as ISI. Whereas in single-carrier systems it can be a serious bottleneck, in OFDM transmission it is possible to add a guard interval, also called a cyclic prefix, between the symbols in order to "absorb" this channel delay spread. The guard interval insertion consists of copying the last samples of the OFDM symbol and pre-pending them to the front of the symbol as illustrated in Figure 1.22. The length of the guard interval is adjusted to the maximum expected delay spread of the channel to allow the receiver to perform the FFT without ISI.

On the other hand, the addition of a guard interval reduces the system data-rate because the time duration of the OFDM symbol is increased while the number of transmitted bits remains the same.

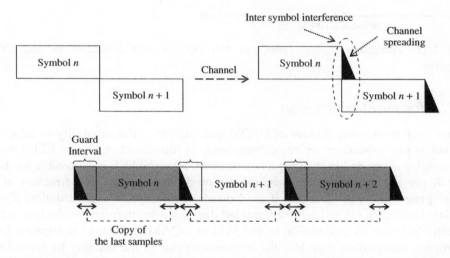

Figure 1.22 Guard interval insertion in order to combat the channel delay spread and the ISI

1.3.6 Windowed OFDM

Recent OFDM transceivers use windowing in transmission in order to reduce out-of-band emissions and to achieve deeper notches around unused subcarriers. The idea is to smooth the transition between OFDM symbols as opposed to the classical rectangular window shape whose frequency response is a sinc $(\sin(f)/f$ function, Equation 1.4), which provides a spectral roll-off of only $-20\,\mathrm{dB}$/decade. Because the windowing reduces the effective guard interval used for combating the ISI, it has to be carefully adjusted for large channel delay spreads.

Figure 1.23 shows how the windowed OFDM shapes the OFDM symbol in the time domain, along with its impact on the power spectral density (PSD).

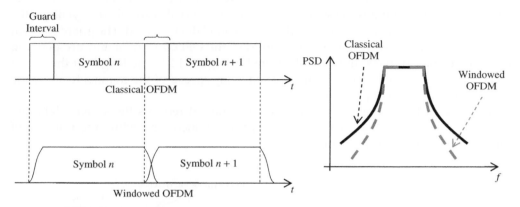

Figure 1.23 Windowed OFDM reduces the out-of-band emissions of the OFDM transmitter

1.3.7 Adaptive Transmission

Another very interesting feature of OFDM modulation is the capability to adapt the modulation per subcarrier, or per sub-channel, as illustrated in Figure 1.24. Because the channel is estimated in the frequency domain, after the FFT, it is possible to obtain the SNR per subcarrier and thus adapt the modulation order as a function of the channel properties. For example, if a 64-quadrature amplitude modulation (QAM) modulated subcarrier is suddenly attenuated due to a deep frequency-selective fading, it is better to lower its modulation to 16-QAM or 4-QAM in order to continue reliably transmitting data, rather than lose the information due to the inability to demodulate

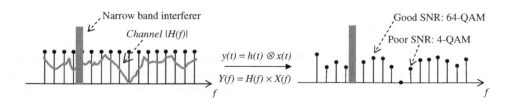

Figure 1.24 Subcarrier adaptive modulation

64-QAM. In addition, the OFDM receiver can detect narrow-band interferers and blank the affected subcarriers.

This technique is only achievable if the receiver is able to send the channel properties to the transmitter such that the transmission can be optimized or adapted to the link quality.

1.3.8 OFDMA for Multiple Access

Whereas in classical OFDM transceivers all the subcarriers of the OFDM symbols are allocated to one user per OFDM symbol, orthogonal frequency division multiple access (OFDMA) divides the subcarriers into sub-channels (Yang, 2010). The concept is to simultaneously share the same OFDM symbol, that is, the channel, between several users at the same time (Figure 1.25). Consequently, in OFDMA all the users receive information about the position of their sub-channel within the OFDM symbol. OFDMA also uses frequency diversity between the users in order to limit the impact of frequency-selective fading which can affect only certain sub-channels, that is, particular subcarriers. OFDMA can resourcefully control the data-rate of each user by adapting the number of allocated subcarriers.

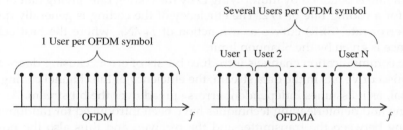

Figure 1.25 In OFDMA the subcarriers of one OFDM symbol are shared between several users

1.3.9 Scalable OFDMA

Because spectrum allocation is imposed by the regulatory rules of each country as well as the network operators, new standards like 3GPP LTE or mobile WiMAX have introduced scalable orthogonal frequency division multiple access (S-OFDMA) in order to be able to customize the OFDMA channel bandwidth (Yang, 2010). However, if the number of subcarriers is not adapted to the channel bandwidth, that is, constant, the system performance will be dependent on the latter. Indeed, narrowing the OFDMA channel bandwidth results in smaller subcarrier spacing and consequently the system becomes more sensitive to Doppler shift, phase noise, and frequency errors, impacting the system specification and complexity. Conversely, if the channel bandwidth is increased, the spectral efficiency lowers because the subcarrier spacing maybe overspecified.

The principle of S-ODFMA is to scale the number of subcarriers with the channel bandwidth in order to keep the subcarrier spacing constant and independent of the system bandwidth.

1.3.10 OFDM DBB Architecture

Figure 1.26 gives an overview of the DBB of a typical OFDM transceiver, which can be decomposed in three separate domains:

- *The bit processing, where the information coming from the MAC is coded and decoded using a forward error correction (FEC) coding.*

The aim of the channel coding, also called FEC, is to add coded redundant bits to the useful bit stream in order to be more robust to channel fading and noise. The channel decoder at reception uses this redundancy for detecting errors and is able to correct them if the SNR is high enough. The coding rate specifies the ratio between the useful bits and the total number of bits that the code generates. For example, a coding rate of $1/2$ means that the channel coding output is 2 bits for every useful bit at the input; that is, we have one redundant bit. If D is the PHY data rate including the coding bits, the useful data rate is obtained by multiplying D by the coding rate, giving half of the PHY data rate for a coding rate of $1/2$. The efficiency of the coding is generally quantified using bit-error-rate (BER) curves as a function of E_b/N_0, where the best achievable performance is given by the Shannon limit.

Because communication channels introduce bursts of errors causing degradation of several consecutive bits in the data stream, the efficiency of the channel coding alone is not optimal, as it assumes independent errors spread over the bit stream. As a result, interleaving and deinterleaving techniques have been introduced for randomizing the bit ordering between the transmitter and the receiver, and thus also the position of errors at the input of the decoder. This allows an improvement in the performance of the channel coding.

- *The frequency-domain processing, where the bits are represented in a complex plane as symbols used to modulate the OFDM subcarriers.*

In transmission, after the channel coding the symbol mapping transforms the bit stream into constellation points. The complexity of the constellation depends on the modulation order. Figure 1.27 shows the mapping for 16-QAM. The constellation points define the magnitudes and the phases of the symbols to apply to the subcarriers before the inverse fast Fourier transform (IFFT) (Figure 1.28).

At the receiver, after the FFT the phase and magnitude of each carrier are extracted and a decision must be taken about which symbol was sent by the transmitter. Due to the channel response each carrier has been distorted in phase and in amplitude, introducing errors in the position of the received constellation points. Consequently, before the symbol demapping, an algorithm is used for estimating the channel response which

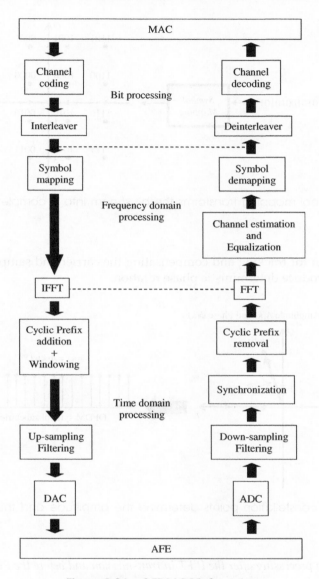

Figure 1.26 OFDM DBB Overview

can be represented in the frequency domain by a phase and an amplitude as a function of subcarrier position. Generally, the channel estimation is performed using training sequences sent by the transmitter or known subcarriers called pilots. Afterwards, the channel estimate is utilized to equalize (i.e., remove the distortion introduced by the channel) all the subcarriers before the symbol demapping, which finally extracts the bit stream from the received equalized constellation. Other algorithms are also used in the

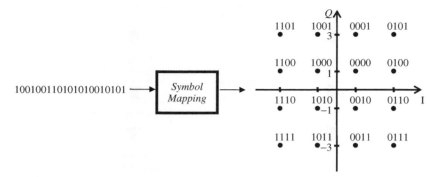

Figure 1.27 Symbol mapping transforms the bit stream into a complex constellation, here 16-QAM

frequency domain for tracking and compensating the carrier and sampling frequency offsets which introduce deterministic phase rotation.

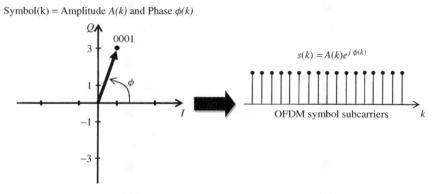

Figure 1.28 The constellation points determine the amplitude and the phase of the subcarriers

- *The time-domain processing after the IFFT in transmission and before the FFT in reception.*

In transmission, after the IFFT the data is now in the time domain where the cyclic prefix is added, which determines the guard interval, and windowing is performed on the OFDM symbol. Finally, up-sampling and digital filters can be used if the DAC clock frequency is higher than the assumed IFFT sampling rate.

In reception, down-sampling and digital filtering can be present if the ADC clock frequency is different than the assumed FFT sampling rate. Time synchronization is required in order to find the precise instants when the OFDM symbols start before applying the FFT. This synchronization can be made, for example, using a known sequence sent by the transmitter or via an autocorrelation using the cyclic prefix and the end of the OFDM symbol.

1.3.11 OFDM-Based Standards

Today, many communication standards use OFDM as modulation for their PHY, notably due to its robustness in the presence of severe channel conditions, as well as its low-complexity equalization done in the frequency domain, which is much simpler than the time-domain equalization used in conventional single-carrier modulation.

The first OFDM-based standards were introduced in the mid 1990s by ETSI (European Telecommunications Standard Institutes) for Digital Audio Broadcasting (DAB) and Digital Video Broadcasting (DVB-T, T for terrestrial). Fifteen years later many others have been developed and are now operational, either for wireless or cable, including:

- Wireless:
  - WiFi (wireless fidelity) for wireless local area networks (WLANs): IEEE 802.11a, g, n.
  - The mobility mode of the IEEE 802.16 wireless metropolitan area networks (WMANs) mobile-WiMAX: IEEE 802.16e based on S-OFDMA.
  - The fourth-generation mobile broadband standard: 3GPP LTE based on S-OFDMA.
  - The terrestrial mobile TV: DVB-H (Digital Video Broadcasting-Handheld).
- Cable:
  - ADSL (asymmetric digital subscriber line) and VDSL (very-high-bit-rate digital subscriber line) for high-definition TV and Internet access over copper phone wires.
  - PLC: Home Plug AV, IEEE 1901, ITU-T G.hn.
  - Multimedia Over Coax Alliance for home networking: MoCA.

1.4 SNR, EVM, and E_b/N_0 Definitions and Relationship

1.4.1 Bit Error Rate

In order to quantify the performance of communication systems, we often find in the literature the use of a BER metric giving the probability of error in terms of the number of corrupted bits per bits received (Proakis, 1995). Depending on the application the BER target can vary from 10^{-3} to 10^{-15}, that is, one erroneous bit every 1000 or 10^{15} bits, respectively. Consequently, the system simulations for measuring these very low BER levels are generally time consuming and cumbersome because large amounts of data, several millions of bits, have to be processed using an accurate model of the DBB chain.

Because the AFE is composed of several blocks (LNA, phase-locked loop (PLL), mixer, etc.) designed by different engineers, it is not practical to directly use the BER as a specification. RF analog designers prefer to work with SNR and/or EVM as performance metrics for specifying and designing the AFE blocks. So it is important that the system designer be able to translate the BER target into SNR or EVM specifications for the AFE development. One technique is to extract the required SNR or EVM from the BER curves, but as they are commonly represented as a function of E_b/N_0, a relationship between these various performance metrics is needed.

Figure 1.29 is a plot of BER curves as a function of E_b/N_0 for four different modulations: 4-QAM, 16-QAM, 64-QAM, and 1024-QAM. If the goal is to achieve a BER lower than 10^{-6}, we clearly see that 1024-QAM requires a much higher E_b/N_0 than 4-QAM, approximately 18 dB higher. With this illustration we can imagine why high data-rate communication systems using high-order modulations need high-performance transceivers, resulting in increased design complexity and increased costs of the final solutions.

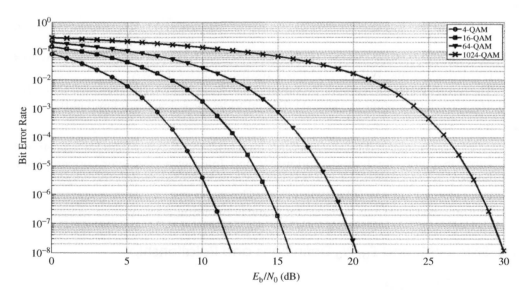

Figure 1.29 BER vs. E_b/N_0 for 4-QAM, 16-QAM, 64-QAM, and 1024-QAM modulations

1.4.2 SNR versus EVM

The *SNR* is defined as the ratio between the received signal power P_s and the noise power P_n integrated within the receiver noise bandwidth B; it is used to predict the performance of the system in terms of minimum receiver sensitivity, or the minimum input signal power which guarantees a certain quality of the communication link.

$$\text{SNR} = \frac{\text{Signal Power}}{\text{Noise Power}} = \frac{P_s}{P_n} = \frac{\frac{1}{T}\int |s(t)|^2 \, dt}{N_0 B} \qquad (1.19)$$

where P_s and P_n are the power of the received signal and the power of the noise in watts, respectively. N_0 is the PSD of the noise in watts per hertz.

In modern communication systems using digital M-ary modulation, the signal is generated using complex symbols:

$$s(t) = x_I(t) + jx_Q(t) \qquad (1.20)$$

where $x_I(t)$ and $x_Q(t)$ are the in-phase and quadrature components of the signal. If we introduce the in-phase and quadrature components of the additive noise, $n_I(t)$ and $n_Q(t)$, respectively, we obtain the noisy signal:

$$s(t) = x_I(t) + jx_Q(t) + n_I(t) + jn_Q(t) \qquad (1.21)$$

and the SNR defined in Equation 1.19 can be expressed as

$$\text{SNR} = \frac{\frac{1}{T}\int \left| x_I^2(t) + x_Q^2(t) \right| \, dt}{\frac{1}{T}\int \left| n_I^2(t) + n_Q^2(t) \right| \, dt} \qquad (1.22)$$

Equation 1.22 shows a direct estimation of the SNR which can be computed in simulation for quantifying the system performance. But in simulation, how can one extract the signal from the noise in Equation 1.21 in order to calculate this ratio? It is not at all obvious, especially for OFDM signals which look like noise; that is, not specified in a deterministic way in the time domain.

One solution is to estimate the EVM, which is another performance metric used by RF analog engineers by measuring the modulation accuracy directly on the constellation points after demodulation (Figure 1.30).

The EVM is defined as the root-mean-square (RMS) of the error between the measured symbols s_n and the ideal ones $s_{0,n}$:

$$\text{EVM} = \sqrt{\frac{\frac{1}{N}\sum_{n=1}^{N} \left| s_n - s_{0,n} \right|^2}{\frac{1}{N}\sum_{n=1}^{N} \left| s_{0,n} \right|^2}} \qquad (1.23)$$

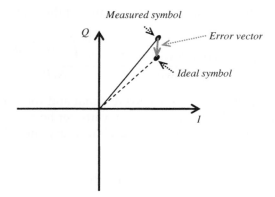

Figure 1.30   Modulation error vector used for estimating the EVM

which can be rewritten:

$$
\text{EVM} = \sqrt{\frac{\frac{1}{N}\sum_{n=1}^{N}\left|x_{\text{I},n} - x_{\text{I0},n}\right|^2 + \left|x_{\text{Q},n} - x_{\text{Q0},n}\right|^2}{\frac{1}{N}\sum_{n=1}^{N}\left|x_{\text{I0},n}\right|^2 + \left|x_{\text{Q0},n}\right|^2}}
\tag{1.24}
$$

where $x_{\text{I},n}$ and $x_{\text{Q},n}$ are the coordinates of the measured symbols in the complex domain, and $x_{\text{I0},n}$ and $x_{\text{Q0},n}$ are those of the ideal ones.

Using Equation 1.21 we can express the noisy symbols as a function of the ideal ones and the quadrature components of an additive noise:

$$
\begin{aligned}
x_{\text{I},n} &= x_{\text{I0},n} + n_{\text{I},n} \\
x_{\text{Q},n} &= x_{\text{Q0},n} + n_{\text{Q},n}
\end{aligned}
\tag{1.25}
$$

Combining Equations 1.24 and 1.25, the EVM becomes

$$
\text{EVM} = \sqrt{\frac{\frac{1}{N}\sum_{n=1}^{N}\left|n_{\text{I},n}\right|^2 + \left|n_{\text{Q},n}\right|^2}{\frac{1}{N}\sum_{n=1}^{N}\left|x_{\text{I0},n}\right|^2 + \left|x_{\text{Q0},n}\right|^2}}
\tag{1.26}
$$

which is the square root of the ratio between the noise power and the signal power; that is, the inverse of the SNR:

$$
\text{EVM} = \sqrt{\frac{\text{Noise Power}}{\text{Signal Power}}} = \frac{1}{\sqrt{\text{SNR}}}
\tag{1.27}
$$

Equation 1.27 shows that the SNR can be directly computed from the value of the EVM:

$$\text{SNR} = \frac{1}{\text{EVM}^2} \tag{1.28}$$

giving in decibels:

$$\text{SNR}_{\text{dB}} = 10 \log_{10}\left(\frac{1}{\text{EVM}^2}\right) = -20 \log_{10}(\text{EVM}) \tag{1.29}$$

1.4.3 SNR versus E_b/N_0

The ratio E_b/N_0, mainly used in BER simulations, quantifies the ratio between the energy per bit and the PSD of the noise.

The energy per bit is defined as the ratio between the signal power and the bit rate:

$$E_b = \frac{P_s}{R_b} \tag{1.30}$$

in which P_s is the signal power in watts and R_b is the data rate in bits/second, giving E_b/N_0 in watt-seconds/bit; that is, joules/bit.

By introducing the noise power, Equation 1.30 can be expressed as a function of the system SNR:

$$E_b = \frac{P_s}{N_0 B} \times \frac{N_0 B}{R_b} = \text{SNR} \times \frac{N_0 B}{R_b} \tag{1.31}$$

in which B is the equivalent noise bandwidth.

Normalizing Equation 1.31 by the noise spectral density N_0, we obtain the formula for E_b/N_0 as a function of the SNR, system bandwidth and bit rate:

$$\frac{E_b}{N_0} = \text{SNR} \times \frac{B}{R_b} \tag{1.32}$$

Combining Equations 1.28 and 1.32, E_b/N_0 can also be expressed as function of the EVM:

$$\frac{E_b}{N_0} = \frac{1}{\text{EVM}^2} \times \frac{B}{R_b} \tag{1.33}$$

Equation 1.32 shows that E_b/N_0 and SNR are equivalent if the system bandwidth is equal to the bit rate, meaning that each transmitted bit requires the full system bandwidth. On the other hand, for communication systems using spread spectrum techniques, like code division multiple access (CDMA) in mobile phones or in GPS, the RF analog noise bandwidth B can be much larger than the information data rate. In practice this means that, after demodulation of the data, each information bit occupies a smaller bandwidth than the RF analog noise bandwidth; the result is a processing gain:

$$\text{PG} = \frac{B}{R_b} \tag{1.34}$$

which can be seen directly as the gain in E_b/N_0 above the SNR in Equation 1.32.

1.4.4 Complex Baseband Representation

In general for system simulations we prefer to work with the complex baseband signal in order to avoid the use of high sampling rates imposed by up-conversion to the carrier frequency, which does not convey any additional information.

The conversion from a real bandpass signal to a complex baseband signal is illustrated in Figure 1.31.

Let us write the original baseband signal at the transmitter input as

$$s(t) = x_I(t) + jx_Q(t) \tag{1.35}$$

The transmitter frequency up-converts the complex baseband signal to a real bandpass signal using a quadrature mixer driven by an LO. The real bandpass signal at the transmitter output, including the carrier frequency, is

$$s_{RF}(t) = \sqrt{2}\,\text{Re}\left\{s(t) \times e^{j2\pi F_c t}\right\} = x_I(t)\sqrt{2}\cos(\omega_c t) - x_Q(t)\sqrt{2}\sin(\omega_c t) \tag{1.36}$$

By assuming an additive noise, the real bandpass signal becomes

$$s_{RF}(t) = x_I(t)\sqrt{2}\cos(\omega_c t) - x_Q(t)\sqrt{2}\sin(\omega_c t) + n_I(t)\sqrt{2}\cos(\omega_c t) + n_Q(t)\sqrt{2}\sin(\omega_c t) \tag{1.37}$$

$n_I(t)$ and $n_Q(t)$ being the quadrature components of the baseband noise.

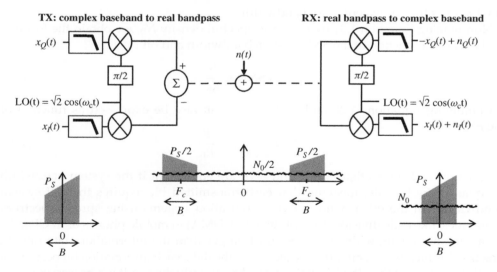

Figure 1.31   From real bandpass to complex baseband used in system simulations

The SNR at RF is estimated by calculating

$$
\mathrm{SNR_{RF}} = \frac{\left\langle \left[x_I(t)\sqrt{2}\cos(\omega_c t) - x_Q(t)\sqrt{2}\sin(\omega_c t) \right]^2 \right\rangle}{\left\langle \left[n_I(t)\sqrt{2}\cos(\omega_c t) + n_Q(t)\sqrt{2}\sin(\omega_c t) \right]^2 \right\rangle} \tag{1.38}
$$

If we only consider the signal of interest around the carrier frequency, that is, ignoring the terms at two times the carrier frequency, we obtain

$$
\mathrm{SNR_{RF}} = \frac{\left\langle x_I^2(t) + x_Q^2(t) \right\rangle}{\left\langle n_I^2(t) + n_Q^2(t) \right\rangle} = \frac{P_s}{N_0 B} \tag{1.39}
$$

in which P_s is the signal power in watts, B is the noise bandwidth in hertz, and N_0 is the noise spectral density in watts/hertz.

The conversion from real bandpass to complex baseband is done in reception by using a complex frequency down-conversion:

$$
\begin{aligned}
s_{BB}(t) = &\left[x_I(t)\sqrt{2}\cos(\omega_c t) - x_Q(t)\sqrt{2}\sin(\omega_c t) \right. \\
&+ \left. n_I(t)\sqrt{2}\cos(\omega_c t) + n_Q(t)\sqrt{2}\sin(\omega_c t) \right] \times \sqrt{2}\cos(\omega_c t) \\
&+ j\left[x_I(t)\sqrt{2}\cos(\omega_c t) - x_Q(t)\sqrt{2}\sin(\omega_c t) \right. \\
&+ \left. n_I(t)\sqrt{2}\cos(\omega_c t) + n_Q(t)\sqrt{2}\sin(\omega_c t) \right] \times \sqrt{2}\sin(\omega_c t)
\end{aligned} \tag{1.40}
$$

After lowpass filtering to remove the high-frequency components around $2\omega_c$, we obtain

$$
s_{BB}(t) = (x_I(t) - jx_Q(t)) + (n_I(t) + jn_Q(t)) \tag{1.41}
$$

The SNR in baseband is then

$$
\mathrm{SNR_{BB}} = \frac{\left\langle x_I^2(t) + x_Q^2(t) \right\rangle}{\left\langle n_I^2(t) + n_Q^2(t) \right\rangle} = \frac{P_s}{N_0 B} \tag{1.42}
$$

which is strictly equal to the RF SNR of the real bandpass signal defined in Equation 1.39, meaning that performance with respect to additive noise is unchanged and the complex baseband model can be used equivalently to the real bandpass model in system simulations.

We will see in the next chapter that the RF impairments are also translated to equivalent models at complex baseband.

References

Abidi, A.A. (1995) Direct-conversion radio transceivers for digital communications. *IEEE Journal on Solid-State Circuits*, **30**, 1399–1410.

Abidi, A. (2000) Wireless transceivers in CMOS IC technology: the new wave, Proceedings of the Symposium on VLSI Technology, pp. 151–158.

Armstrong, J. (2007) *OFDM*, John Wiley & Sons, Inc.

Bingham, J.A.C. (1990) Multicarrier modulation for data transmission: an idea whose time has come. *IEEE Communications Magazine*, **28**, 5–14.

Brandolini, M., Rossi, P., Manstretta, D., and Svelto, F. (2005) Toward multi-standard mobile terminals – fully integrated receivers requirements and architectures. *IEEE Transactions on Microwave Theory and Techniques*, **53** (3), 1026–1038.

Crols, J. and Steyaert, M.S.J. (1998) Low-if topologies for high-performance analog front ends of fully integrated receivers. *IEEE Transactions on Circuits Systems II*, **45**, 269–282.

Farrukh, F., Baig, S., and Mughal, M.J., (2009) MMSE equalization for discrete wavelet packet based OFDM. Proceedings of IEEE International Conference on Electrical Engineering.

Fettweis, G., Löhning, M., Petrovic, D. *et al.* (2007) Dirty RF: a new paradigm. *International Journal of Wireless Information Networks*, **14** (2), 133–148.

Haykin, S. (2001) *Communication Systems*, 4th edn, John Wiley & Sons, Inc.

Hoeher, P., Kaiser, S., and Robertson, P. (1997) Pilot-symbol aided channel estimation in time and frequency. Proceedings of Globecom, pp. 90–96.

Iwai, H. (2000) CMOS technology for RF applications. *Proceedings of the 22nd International Conference on Microelectronics*, Vol. 1, pp. 27–34.

Karp, T., Trautmann, S., and Fliege, N.J. (2002) Frequency domain equalization of DMT/OFDM systems with insufficient guard interval. Proceedings of the IEEE International Conference on Communications, pp. 1646–1650.

Karp, T., Wolf, M., Trautmann, S., and Fliege, N.J. (2003) Zero-forcing frequency domain equalization for DMT systems with insufficient guard interval. Proceedings of IEEE International Conference on Acoustics, Speech, and Signal Processing, pp. 221–224.

Morelli, M. and Pun, M. (2007) Synchronization techniques for orthogonal frequency division multiple access (OFDMA) a tutorial review. *Proceedings of the IEEE*, **95** (7), 1394–1427.

Lee, T.H. (1998) *The Design of CMOS Radio-Frequency Integrated Circuits*, Cambridge University Press.

Li, Y.G. and Stuber, G.L. (2006) *Orthogonal Frequency Division Multiplexing for Wireless Channels*, 1st edn, Springer.

Prasad, R. (2004) *OFDM for Wireless Communications Systems*, Artech House.

Proakis, J.G. (1995) *Digital Communications*, McGraw-Hill, New York.

Rappaport, T.S. (1996) *Wireless Communications: Principles and Practice*, Prentice Hall.

Razavi, B. (1997) Design considerations for direct conversion receivers. *IEEE Transactions on Circuits and Systems II: Analog and Digital Signal Processing*, **44**, 428–435.

Razavi, B. (1998a) *RF Microelectronics*, Prentice Hall.

Razavi, B. (1998b) Architecture and circuits for RF CMOS receivers. Proceedings IEEE Custom Integrated Circuits Conference, pp. 393–400.

Steele, R. (1995) *Mobile Radio Communications*, IEEE Press.

Tucker, D.G. (1954) The history of the homodyne and the synchrodyne. *Journal of the British Institution of Radio Engineers*, **14**, 143–154.

Tufvesson, F. and Maseng T. (1997) Pilot assisted channel estimation for OFDM in mobile cellular systems. Proceedings of Vehicular Technology Conference, pp. 1639–1643.

Van de Beek, J.J., Edfors, O., Sandell, M. *et al.* (1995) On channel estimation in OFDM systems. Proceedings of Vehicular Technology Conference, pp. 815–819.

Van Nee, R.D.J. and Prasad, P. (2000) *OFDM for Wireless Multimedia Communications*, Artech House.

Weinstein, S.B. and Ebert, P.M. (1971) Data transmission by frequency-division multiplexing using the discrete Fourier transform. *IEEE Transactions on Communications*, **19**, 628–634.

Yang, S.C. (2010) *OFDMA System Analysis and Design*, Mobile Communication Series, Artech House.

2

RF Analog Impairments Description and Modeling

2.1 Introduction

Modern on-chip wireless transceivers usually use zero-IF, also called direct conversion or low-IF architectures because both are well suited to system integration by avoiding the use of external RF filters as are seen in the superheterodyne case.

Figure 2.1 presents a simple schematic overview of zero-IF and low-IF AFE architectures. The RF analog receiver is composed of a LNA fixing its sensitivity, a LO and a quadrature mixer for frequency down-converting the signal around DC (zero-IF) or in low-IF, and finally anti-alias filtering and ADC for sampling and digitizing the baseband signal. The RF analog transmitter is composed of a DAC with reconstruction filtering for converting the baseband digital signal into analog, a LO and a quadrature mixer for frequency up-converting the signal to RF, and finally a PA for delivering the required transmit power.

In this chapter we will especially focus on the principal impairments present in the RF AFE:

- thermal and flicker noises;
- LO phase noise;
- sampling jitter;
- carrier frequency offset (CFO) and sampling frequency offset (SFO);
- DAC and ADC quantization noise and clipping;
- quadrature imbalance;
- second-order intercept point (IP2)/third-order intercept point (IP3) nonlinearities and PA distortions.

RF Analog Impairments Modeling for Communication Systems Simulation: Application to OFDM-based Transceivers, First Edition. Lydi Smaini.
© 2012 John Wiley & Sons, Ltd. Published 2012 by John Wiley & Sons, Ltd.

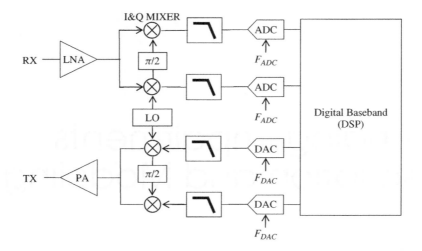

Figure 2.1   Zero-IF or low-IF transceiver architecture

From the description of these imperfections we derive mathematical models which can be easily incorporated into any system simulation chain for studying their effects on transmitter and receiver performances. Furthermore, for illustrating the theory, we study the impact of these impairments on OFDM signal EVM/SNR and constellations.

2.2 Thermal Noise

2.2.1 Additive White Gaussian Noise

In communication systems theory and simulations we often introduce an additive noise $n(t)$ for studying the transceiver performances, as shown in Figure 2.2; either we use its PSD (watts/hertz) when analyzing E_b/N_0 or its power (watts) when studying SNR or EVM. Most of the time this noise is assumed additive, white, and Gaussian noise (AWGN).

Why this assumption? In practice the AFE of any real communication system is affected by random fluctuations of the charges (electrons) present in any physical

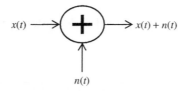

Figure 2.2   Additive noise introduced in simulations

device. These fluctuations are known as thermal noise and are dependent on the temperature, as shown next (Davenport and William, 1958; Ott, 1976). For example, in Figure 2.3 a simple passive resistor R generates an RMS noise voltage equal to

$$\sqrt{\langle v_{n}^{2}(t) \rangle} = v_{n} = \sqrt{4kTRB} \tag{2.1}$$

where the angle brackets denote the average operator, $k = 1.38 \times 10^{-23}$ J/K is Boltzmann's constant, T is the temperature in kelvin, R is the value of the resistor in ohms, and B is the integration/measurement bandwidth in hertz.

For example, a $50\,\Omega$ resistor and $20\,\text{MHz}$ bandwidth will generate a noise voltage of $4\,\mu\text{V}$ RMS at room temperature (290 K).

Another representation of the noisy resistor is to use the Thevenin model of a noise-free resistor in series with a noise voltage source (Figure 2.4) having the RMS value of Equation 2.1.

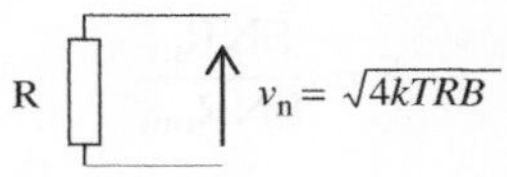

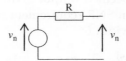

Figure 2.3 Noise voltage generated by a passive resistor

If the noisy resistor is the source impedance (antenna) of a receiver or a load (antenna) connected to a transmitter (Figure 2.5), the maximum power is delivered when the source and the load are matched in impedance. In this case the noise voltage across the antenna is

$$\frac{v_{n}}{2} = \sqrt{kTRB} \tag{2.2}$$

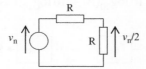

Figure 2.4 Noisy resistor equivalent model: voltage source with a noise-free resistor in series

Figure 2.5 Thermal noise in a perfect matched circuit

giving a noise power

$$P_n = \frac{v_n^2/4}{R} = kTB \tag{2.3}$$

If we normalize the thermal noise equation (2.3) by the system bandwidth B, we obtain the well known thermal noise density N_0 in watts/hertz:

$$N_0 = kT \tag{2.4}$$

At room temperature, $T = 290\,\text{K}$, we obtain $N_0 \approx -174\,\text{dBm/Hz}$.

2.2.2 Noise Figure and Sensitivity

The impact of the thermal noise in communication systems is quantified using an RF analog term called noise figure which is defined as the ratio between the SNR at the input and the SNR at output of a device (Friis, 1944):

$$F = \frac{\text{SNR}_{\text{in}}}{\text{SNR}_{\text{out}}} \tag{2.5}$$

Because any realizable AFE will degrade the SNR due to its intrinsic noise, the noise figure is always greater than one; that is, positive in decibels. Transceivers are generally composed of cascaded blocks (Figure 2.6), for example, several stages of amplification in reception; the overall noise figure can be derived from the Friis equation:

$$F_{\text{RX}} = F_1 + \frac{F_2 - 1}{G_1} + \frac{F_3 - 1}{G_1 G_2} \cdots + \frac{F_N - 1}{G_1 G_2 \cdots G_{N-1}} \tag{2.6}$$

where F_1 and G_1 are the noise figure and the gain of the first stage, respectively, and F_N and G_N are those of the last one.

From Equation 2.6 one can conclude that the first stage of a receiver needs a very low noise figure and high gain in order to guarantee good sensitivity; that is, the minimum signal level at the input for a specified output SNR. The receiver sensitivity can be easily derived from the minimum SNR requirement in the presence of thermal noise integrated over the system bandwidth B:

$$\text{SNR}_{\text{min}} = \frac{\text{Sensi}}{F_{\text{RX}} \times kTB} \Rightarrow \text{Sensi} = F_{\text{RX}} \times kTB \times \text{SNR}_{\text{min}} \tag{2.7}$$

which is typically expressed in dBm assuming 290 K:

$$\text{Sensi(dBm)} = -174 + 10 \log_{10}(B) + 10 \log_{10}(F_{\text{RX}}) + 10 \log_{10}(\text{SNR}_{\text{min}}) \tag{2.8}$$

Figure 2.6 Cascaded noisy blocks degrading the SNR

2.2.3 Cascaded Noise Voltage in IC Design

In IC design, with the exception of the antenna, which is off chip and generally matched to the LNA or the PA (typically 50 Ω) for maximizing the transfer of power, there is no matching anymore on-chip and thus IC designers prefer to work with noise voltage, or noise voltage density, instead of noise figure defined in the previous section.

The noise voltage density is simply the noise voltage divided by the square root of the system bandwidth. For example, the noise density generated by a resistor R is Equation 2.1 divided by $\sqrt{B}$:

$$V_n = \frac{v_n}{\sqrt{B}} = \frac{\sqrt{4kTRB}}{\sqrt{B}} = \sqrt{4kTR} \tag{2.9}$$

where V_n is the RMS noise voltage density in $V/\sqrt{Hz}$.

Figure 2.7 presents unmatched cascaded blocks in which the noise contribution of each block is not defined by a noise figure but rather by its input-referred noise voltage. If we consider that the noise contributions of all the blocks are uncorrelated, the total output noise voltage can be estimated using a cascaded quadratic sum:

$$v_{n\,OUT}^2 = G_1 G_2 \cdots G_N (v_n^2 + v_{n1}^2) + G_2 \cdots G_N (v_{n2}^2) \cdots + G_N v_{nN}^2 \tag{2.10}$$

By normalizing the total output noise voltage by the overall gain, we obtain the total input-referred noise voltage:

$$v_{n\,IN}^2 = \frac{v_{n\,OUT}^2}{G_{RX}} = \frac{v_{n\,OUT}^2}{G_1 G_2 \cdots G_N} = v_n^2 + v_{n\,RX}^2 \tag{2.11}$$

in which G_{RX} is receiver power gain, and $v_{n\,RX}$ is the intrinsic receiver input-referred noise voltage:

$$v_{n\,RX}^2 = v_{n1}^2 + \frac{v_{n2}^2}{G_1} + \cdots + \frac{v_{nN}^2}{G_1 G_2 \cdots G_{N-1}} \tag{2.12}$$

where v_{n1} and G_1 are the RMS noise voltage (integrated over the system bandwidth B) and the gain of the first block, respectively, and v_{nN} and G_{nN} are those of the last one. So the receiver noise figure can be derived by applying the following ratio:

$$F_{RX} = \frac{v_{n\,OUT}^2}{G_{RX} v_n^2} = \frac{v_n^2 + v_{n\,RX}^2}{v_n^2} = 1 + \frac{v_{n\,RX}^2}{4kTRB} \tag{2.13}$$

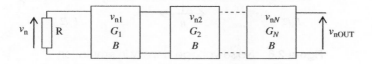

Figure 2.7 Unmatched cascaded blocks

2.2.4 AWGN in Simulations

Usually simulation tools generate AWGN using a function returning normally distributed pseudorandom numbers with an RMS value equal to one. But for practical system simulations the level of this noise has to be adjusted to the real value of the thermal noise taking into account the simulation bandwidth corresponding to the sampling frequency F_s, which is generally much higher than the bandwidth of interest B.

In simulations the AWGN is spread over all the Nyquist band $\pm F_s/2$; consequently, if we consider an RMS noise voltage v_n integrated over a bandwidth B, such as the noise voltage of a block defined in the previous section, for accurate simulations we have to generate an RMS noise voltage equal to

$$v_{n\,\text{SIMU}} = V_n\sqrt{F_s} = \frac{v_n}{\sqrt{B}}\sqrt{F_s} \qquad (2.14)$$

where V_n is the real RMS noise voltage density in $\text{V}/\sqrt{\text{Hz}}$, B is the system bandwidth, and F_s is the simulation sampling frequency.

Equation 2.14 is illustrated in Figure 2.8.

In the classical case of thermal noise power $P_n = kTB$ that one encounters in many references for simulating the performance of a matched receiver as a function of the SNR or E_b/N_0, we have to generate an AWGN with a power equal to

$$P_{n\,\text{SIMU}} = \frac{v_{n\,\text{SIMU}}^2}{R} = kTB\frac{F_s}{B} = N_0 F_s \qquad (2.15)$$

giving an RMS noise voltage at receiver input of

$$v_{n\,\text{SIMU}} = \sqrt{RN_0 F_s} \qquad (2.16)$$

where R is the receiver input impedance.

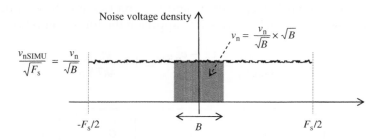

Figure 2.8   Noise voltage density in simulation

2.2.5 Flicker Noise and AWGN Modeling

In CMOS transceivers, in addition to the thermal noise we must also take into account the flicker noise, also called $1/f$ noise ("one over f noise"), coming from the transistors (actually active devices) (Kleinpenning, 1990; Vandamme, 1990). This can be a real performance bottleneck for homodyne architecture receivers because it is located around DC, but can be avoided in low-IF architectures (Figure 2.9).

Two main parameters describe the flicker noise: the corner frequency and the noise floor level. In fact, the corner frequency is simply defined as the frequency where the flicker noise ($1/f$ shape, $-10\,\text{dB/decade}$) is equal to the thermal noise floor level (Figure 2.10).

The total noise PSD including the flicker noise and AWGN can be expressed as

$$\text{PSD}_{\text{noise}}(f) = N_0 \left(1 + \frac{f_{\text{corner}}}{f} \right) \tag{2.17}$$

when $f = f_{\text{corner}}$, $\text{PSD}_{\text{noise}}(f_{\text{corner}}) = 2N_0$ (i.e., $3\,\text{dB}$ more than AWGN PSD level N_0).

Knowing the noise PSD, it is possible to obtain the temporal noise $n(t)$ by defining a phase spectrum. In addition, because the noise PSD is only represented for positive frequencies, we have to recreate the spectrum for "negative" frequencies before applying an IFT to generate the temporal waveform.

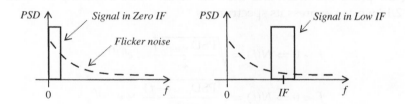

Figure 2.9 Low-IF architecture for avoiding the flicker noise

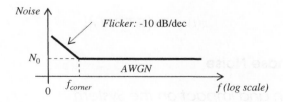

Figure 2.10 PSD of the noise with flicker

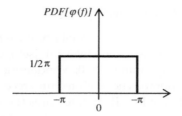

Figure 2.11   PDF of the flicker noise phase spectrum

For a real signal we know that the spectrum is Hermitian:

$$f > 0 \Rightarrow \mathrm{FT}\,[n(t)] = N(f)$$
$$f < 0 \Rightarrow \mathrm{FT}\,[n(t)] = [N(-f)]^* \tag{2.18}$$
$$f = 0 \Rightarrow N(0) = \mathrm{DC}$$

where the asterisk indicates the conjugate operator and FT is the Fourier transform.

Since the flicker noise PSD profile does not carry any phase information, we introduce a phase spectrum $\varphi(f)$ as a random process having a uniform distribution between $-\pi$ and $+\pi$ (Figure 2.11).

Having the phase spectrum and the PSD of the noise including the flicker, using Equation 2.18 we can express its spectrum as

$$f > 0 \Rightarrow N(f) = \sqrt{\frac{\mathrm{PSD}_{\mathrm{noise}}(f)}{2}} \times \mathrm{e}^{j\varphi(f)}$$
$$f < 0 \Rightarrow N(f) = \sqrt{\frac{\mathrm{PSD}_{\mathrm{noise}}(-f)}{2}} \times \mathrm{e}^{-j\varphi(f)} \tag{2.19}$$
$$f = 0 \Rightarrow N(0) = \mathrm{DC}$$

allowing the generation of the temporal noise via an IFT:

$$n(t) = \mathrm{Re}\left\{\mathrm{IFT}\,[N(f)]\right\} \tag{2.20}$$

2.3 Oscillator Phase Noise

2.3.1 Description and Impact on the System

In RF transceivers the complex baseband signal $x(t)$ containing the useful information is up-converted around a carrier frequency f_c and frequency down-converted to baseband, in transmission and reception respectively, using a mixer and an LO generally implemented by a PLL (Gardner, 1979). Without phase noise, the RF signal $y(t)$ can be simply written as

$$y(t) = x(t)\, e^{j2\pi f_c t} \tag{2.21}$$

Now by taking into account the LO phase noise $\theta(t)$, in radians, Equation 2.21 becomes

$$y(t) = x(t)\, e^{j(2\pi f_c t + \theta(t))} = x(t)\, e^{j\theta(t)}\, e^{j2\pi f_c t} \tag{2.22}$$

Like the flicker noise described in Section 2.2.5, the phase noise $\theta(t)$ is commonly described in the frequency domain by its PSD in dBc/Hz (or rad^2/Hz in linear). It is the ratio between the noise power measured in 1 Hz bandwidth, at a frequency offset f_m, and the power of the carrier (Hajimiri and Lee, 1998; Lee and Hajimiri, 2000). It is illustrated in Figure 2.12; ideal oscillators are merely represented by a Dirac function in the frequency domain, whereas real ones present a kind of "skirt" due to the phase noise profile.

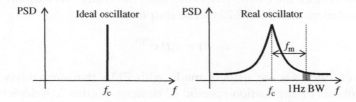

Figure 2.12 Ideal vs. real oscillator degraded by phase noise

Because the output of the LO and the signal are multiplied in the time domain through a mixer, the result is a convolution of their spectrum in the frequency domain:

$$x(t)\, e^{j(2\pi f_c t + \theta(t))} \xleftarrow{\text{Fourier}} X(f - f_c) \otimes \mathrm{FT}\left[e^{j\theta(t)}\right] \tag{2.23}$$

where $\otimes$ represents the convolution operator and FT is the Fourier transform. The impact of the LO phase noise in transceivers is illustrated in Figure 2.13. The LO phase noise will affect the signal by adding an in-band noise scaled by the signal power; it is similar in reception and transmission, but in reception we have to take into account another effect introduced by blockers or interferers called reciprocal mixing. Indeed, because interferers are also "contaminated" by the receiver phase noise, similar to the signal, they can reinject into the signal bandwidth phase noise centered at a frequency offset (defined by the distance between the interferer and the signal) which can be much larger than the receiver bandwidth.

2.3.2 Phase Noise Modeling in the Frequency Domain

Actually, in simulations, it is possible to directly apply the phase noise to the baseband signal for considerably reducing the sampling resolution imposed by modeling at the

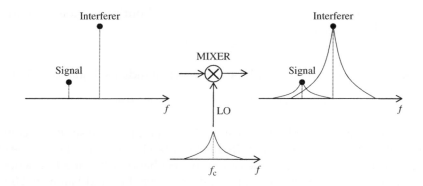

Figure 2.13   Effect of the LO phase noise through the mixer

carrier frequency. Since the carrier frequency does not really carry useful information, for system simulations Equation 2.22 can be simplified:

$$y_{BB}(t) = x(t)\, e^{j\theta(t)} \tag{2.24}$$

In transceivers the LOs are usually made with PLLs that are widely utilized for frequency synthesis in application-specific IC designs. Figure 2.14 depicts the simple diagram of a PLL. It is a closed-loop system composed of a PFD (phase frequency discriminator) comparing the phase and the frequency of the output, generated by a VCO (voltage-controlled oscillator), to the phase and the frequency of a reference oscillator. The error signal, output of the PFD, is low-pass filtered for controlling the VCO in order to track the instantaneous phase of reference oscillator. A frequency divider is commonly present in the feedback path for controlling the output frequency:

$$F_0 = RF_{ref} \tag{2.25}$$

Some definitions of the terms used in PLL design (Gardner, 1979):

- F_{ref} is the reference frequency;
- F_0 is the output frequency;
- R is the feedback divider defining the multiplying factor ($R = F_0/F_{ref}$);

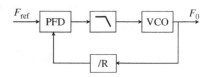

Figure 2.14   Simple PLL scheme used in RF communication systems

- $T_0 = 1/F_0$ is the mean period;
- ω_n is the natural pulsation (related to the PLL bandwidth);
- ξ is the damping coefficient.

If we assume that the PLL loop filter is first order, the closed-loop transfer function of the PLL from input (reference frequency) to output (VCO output frequency) is a second-order low-pass filter with a low frequency gain of R (Gardner, 1979), in the Laplace domain:

$$H_{\text{ref}}(s) = \frac{R(2\xi\omega_n s + \omega_n^2)}{s^2 + 2\xi\omega_n s + \omega_n^2} \tag{2.26}$$

Because the VCO is not an ideal oscillator and then will have phase noise, the closed-loop transfer function of the PLL from VCO added noise is a second-order high-pass filter with a gain of one (Gardner, 1979):

$$H_{\text{VCO}}(s) = \frac{s^2}{s^2 + 2\xi\omega_n s + \omega_n^2} \tag{2.27}$$

We see from these two transfer functions which are asymptotically plotted in Figure 2.15 that the PLL in-band phase noise is especially dominated by the reference oscillator, the PFD and the feedback divider, while the PLL out-of-band noise is mainly a function of the VCO.

In RF analog design the PLL performances are generally defined by their output phase noise profile, which is used by designers to optimize the different parameters of the PLL, and consequently system designers have to know how to properly simulate and specify phase noise requirements.

PLL phase noise can be well specified by system designers using the model depicted in Figure 2.16, the double-side-band (DSB) PSD which is 3 dB greater than the single-side-band (SSB) phase noise profile. It takes into account the flicker noise generated by the PFD, the $1/f$ component, the $1/f^2$ component of the VCO, and two

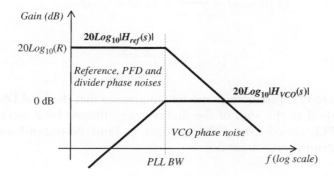

Figure 2.15 PLL transfer functions: low pass for the reference, high-pass for the VCO

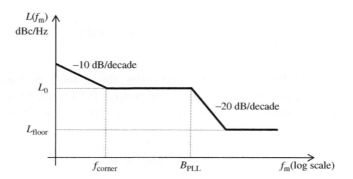

Figure 2.16   PLL double-side band (DSB) phase noise profile in dBc/Hz

noise floors: in-band from the reference/PFD/feedback divider and out-of-band from the VCO.

The $1/f^2$ component can be modeled as a Lorentzian spectrum:

$$L_{1/f^2}(f_m) = \frac{B_{PLL}^2 L_0}{B_{PLL}^2 + f_m^2} \tag{2.28}$$

in which f_m is the frequency offset from the carrier, B_{PLL} is the PLL $-3\,$dB bandwidth, and L_0 is the in-band phase noise level in rad^2/Hz.

Regarding the flicker noise, $1/f$ component, we can model it with the following equation:

$$L_{1/f}(f_m) = \frac{\alpha}{f_m} \tag{2.29}$$

In order to find the constant α, we fix the flicker corner frequency f_{corner} and the PLL phase noise in-band level L_0:

$$L_{1/f}(f_{corner}) = \frac{\alpha}{f_{corner}} = L_0 \Rightarrow \alpha = L_0 f_{corner} \tag{2.30}$$

giving

$$L_{1/f}(f_m) = L_0 \frac{f_{corner}}{f_m} \tag{2.31}$$

In our complete PLL phase noise model we assume that the total DSB phase noise profile is expressed as the sum of the flicker noise, filtered by a second-order filter modeling the PLL closed-loop filtering so as to limit its out-of-band impact, the Lorentzian spectrum, and a noise floor:

$$L(f_m) = L_{1/f}(f_m) \frac{B_{PLL}^2}{B_{PLL}^2 + f_m^2} + L_{1/f^2}(f_m) + L_{floor} \tag{2.32}$$

By rearranging Equation 2.32 we obtain the final PLL phase noise model:

$$L(f_m) = \frac{B_{PLL}^2 L_0}{B_{PLL}^2 + f_m^2}\left(1 + \frac{f_{corner}}{f_m}\right) + L_{floor} \qquad (2.33)$$

In the previous section we assumed a typical PLL phase noise profile, defined by Equation 2.33, but of course this frequency-domain modeling method can be extrapolated to any phase noise profile. We will see in the next section that this frequency modeling can be used to generate the phase noise in the temporal domain, more convenient for simulations, by adding a phase spectrum as for the flicker noise generation described in Section 2.2.5.

Figure 2.17 presents an example of phase noise profile $L(f_m)$ generated using MATLAB® with $L_0 = -95\,$dBc/Hz, $L_{floor} = -150\,$dBc/Hz, $f_{corner} = 1\,$kHz, and $B_{PLL} = 100\,$kHz.

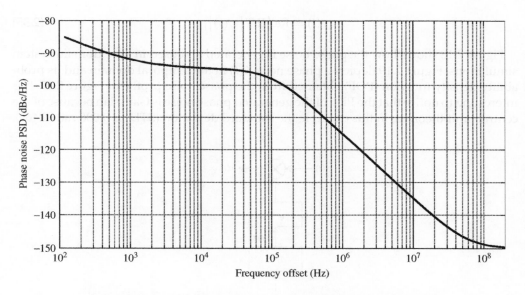

Figure 2.17 Example of a DSB phase noise profile in dBc/Hz

2.3.3 Simulation in Temporal Domain

As we have seen in Section 2.3.2, the LO phase noise is generally specified in the frequency domain in terms of its PSD in dBc/Hz, but in practice it is better to use a temporal-domain model for simulations. Indeed, because the LO and the signal are multiplied in the time domain through a mixer, the result is a convolution of their spectra in the frequency domain, which is much more resource intensive for system simulations.

Consequently, it is preferable to execute the simulation with the temporal representation of the phase noise in order to avoid the convolution operator. As the flicker noise defined in the frequency domain (Section 2.2.5), having only the phase noise PSD $L(f)$ defined by Equation 2.33, we recreate the phase noise spectrum $O(f)$ by introducing a phase spectrum $\varphi(f)$ as a random process with a uniform distribution between $-\pi$ and $+\pi$ (Figure 2.11):

$$f > 0 \Rightarrow O(f) = \sqrt{\frac{L(f)}{2}} \, e^{j\varphi(f)}$$

$$f < 0 \Rightarrow O(f) = \sqrt{\frac{L(-f)}{2}} \, e^{-j\varphi(f)} \tag{2.34}$$

$$f = 0 \Rightarrow O(0) = 1$$

Having the spectrum of the phase noise $O(f)$ specified by Equation 2.34, its temporal-domain representation is obtained via an IFT:

$$\theta(t) = \text{Re} \left\{ \text{IFT} \left[O(f) \right] \right\} \tag{2.35}$$

Figure 2.18 shows how the phase noise is generated and used in time-domain system simulations. Figure 2.19 depicts a phase noise $\theta(t)$ in the temporal domain and its probability density function (PDF) obtained from the phase noise profile of Figure 2.17. It is interesting to notice that the PDF of the generated phase noise is Gaussian because of the central limit theorem; in this example the standard deviation (STD) is 0.008 rad (0.46°).

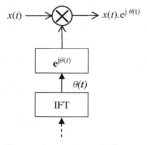

Figure 2.18 Phase noise model used for simulations

2.3.4 SNR Limitation due to the Phase Noise

Let us suppose a noiseless symbol

$$s = x_I + jx_Q \tag{2.36}$$

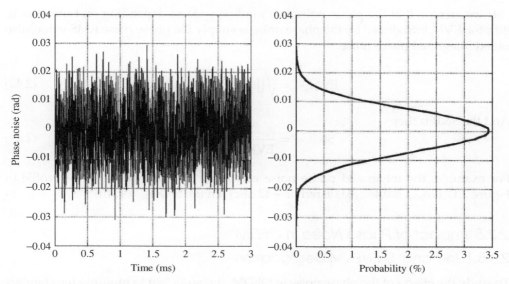

Figure 2.19   Phase noise $\theta(t)$ in the temporal domain generated from phase noise profile of Figure 2.17

Introducing a phase error θ, in radians, the symbol becomes

$$s_{PN} = s\,e^{j\theta} = (x_I + jx_Q)\,e^{j\theta} = x_{IPN} + jx_{QPN} \qquad (2.37)$$

giving an EVM

$$EVM_{PN} = \sqrt{\frac{(x_{IPN} - x_I)^2 + (x_{QPN} - x_Q)^2}{x_I^2 + x_Q^2}} \qquad (2.38)$$

By supposing that $\theta \ll 1$, a valid assumption for LOs used in modern communication transceivers, we can approximate Equation 2.37:

$$s_{PN} \approx (x_I + jx_Q)(1 + j\theta) = (x_I - \theta x_Q) + j(\theta x_I + x_Q) \qquad (2.39)$$

Consequently, the EVM introduced by phase noise is

$$EVM_{PN} = \sqrt{\frac{(x_I - \theta x_Q - x_I)^2 + (\theta x_I + x_Q - x_Q)^2}{x_I^2 + x_Q^2}} \qquad (2.40)$$

and can be simplified as

$$EVM_{PN} = \sqrt{\frac{(x_I^2 + x_Q^2)\theta^2}{x_I^2 + x_Q^2}} = \theta \qquad (2.41)$$

If we assume that the phase noise is a random process depending on time, that is, $\theta(t)$, the EVM introduced by the phase noise is simply the phase noise RMS value, also called integrated phase noise:

$$\text{EVM}_{\text{PN}} = \sqrt{\left\langle |\theta(t)|^2 \right\rangle} = \theta_{RMS} \tag{2.42}$$

And the SNR is then

$$\text{SNR}_{\text{PN}} = \frac{1}{\text{EVM}_{\text{PN}}^2} = \frac{1}{\theta_{RMS}^2} \tag{2.43}$$

For example, the integrated phase noise estimated from the phase noise profile of Figure 2.17 is $\theta_{\text{rms}} = 0.46°$ (0.008 rad or -42 dBc), giving $\text{SNR}_{\text{PN}} = 42$ dB.

2.3.5 Impact of Phase Noise in OFDM

2.3.5.1 In-band: SNR/EVM Degradation

To study the effects of the phase noise in OFDM, it is sufficient to multiply the complex OFDM baseband signal $x(t)$ by $e^{j\theta(t)}$ as indicated by Equation 2.24. By supposing that $\theta(t) \ll 1$, a valid assumption for LOs used in modern communication transceivers, we can write

$$y(t) = x(t)\, e^{j\theta(t)} \approx x(t)(1 + j\theta(t)) \tag{2.44}$$

The OFDM receiver will apply a DFT on the sampled signal $y(n)$ defined in Equation 2.44:

$$Y(m) = \frac{1}{N} \sum_{n=0}^{N-1} y(n)\, e^{-j2\pi(mn/N)} \approx \frac{1}{N} \sum_{n=0}^{N-1} x(n)\, e^{-j2\pi(mn/N)} + \frac{j}{N} \sum_{n=0}^{N-1} x(n)\theta(n)\, e^{-j2\pi(mn/N)} \tag{2.45}$$

where the discrete signal spectrum $x(n)$ can be expressed as an IFT:

$$x(n) = \sum_{k=-N/2}^{N/2-1} s_k\, e^{j2\pi(kn/N)} \tag{2.46}$$

in which N is the total number of subcarriers within the OFDM symbol (including DC and guard bands) and s_k is the complex symbol modulating subcarrier k.

The received signal is composed of the original one, $X(m)$, and an error term $E(m)$ affecting all the subcarriers due to the phase noise:

$$Y(m) \approx X(m) + E(m) \tag{2.47}$$

with

$$X(m) = \frac{1}{N} \sum_{n=0}^{N-1} x(n) \, e^{-j2\pi \frac{mn}{N}}$$

(2.48)

$$E(m) = \frac{j}{N} \sum_{k=-N/2}^{N/2-1} s_k \sum_{n=0}^{N-1} \theta(n) \, e^{j\frac{2\pi}{N}(k-m)n}$$

The error expression $E(m)$ can be decomposed into two terms (Armada, 2001):

$$E(m) = E_{m=k}(m) + E_{m \neq k}(m)$$

(2.49)

- Common phase error (CPE) for $m = k$

$$E_{m=k}(k) = s_k \, \frac{j}{N} \sum_{n=0}^{N-1} \theta(n) = s_k \times j\theta_{\text{mean}}$$

(2.50)

In this case all the subcarriers are biased by the same amount given by the mean of the phase noise. This bias is called CPE because it affects all the subcarriers through a phase rotation of each symbol s_k resulting in a rotation of the entire constellation.

- ICI for $m \neq k$

$$E_{m \neq k}(m) = \frac{j}{N} \sum_{\substack{k=-N/2 \\ m \neq k}}^{N/2-1} s_k \sum_{n=0}^{N-1} \theta(n) \, e^{j(2\pi/N)(k-m)n}$$

(2.51)

which can be rewritten as

$$E_{m \neq k}(m) = \frac{j}{N} \sum_{\substack{k=-N/2 \\ m \neq k}}^{N/2-1} s_k O(m-k)$$

(2.52)

in which $O(m)$ is the discrete phase noise spectrum. Equation 2.52 can be written as a discrete convolution product in the frequency domain between the symbols s_m and the phase noise spectrum $O(m)$.

$$E_{m \neq k}(m) = \frac{j}{N} s_m \otimes O(m)$$

(2.53)

Because the DFT is applied on the OFDM symbol composed of N samples, the result can be seen as a circular convolution of the received signal spectrum (OFDM symbol) by a "$\sin(f)/f$" weighting function (for a rectangular window) having a $-3\,\text{dB}$ main lobe width roughly equal to subcarrier spacing Δf, as depicted in Figure 2.20. Due to the phase noise, each subcarrier will be affected by the other $N-1$ subcarriers, causing ICI, and the impact is a loss of orthogonality between the subcarriers.

ICI can be seen as Gaussian noise (if $N \gg 1$, by the central limit theorem) because it is the sum of $N-1$ modulated subcarriers weighted by the phase noise frequency shifted by $(m-k)$. The bandwidth B_{PLL} of the phase noise, compared with the subcarrier spacing Δf, is very important for differentiating CPE and ICI.

The impact of the phase noise bandwidth B_{PLL} on the SNR/EVM can be expressed as follows:

- If $\Delta f/2 > B_{\text{PLL}}$, that is, half of the subcarrier spacing larger than phase noise bandwidth, the major part of the phase noise power being in the main lobe of the $\sin(f)/f$, the SNR/EVM is primarily affected by the CPE, which can be corrected in the DBB by estimating the rotation of the constellation. In this case, the ICI is not predominant. In the time domain this corresponds to phase noise variations which are slower than the OFDM symbol time ($\sim 1/\Delta f$) and can be tracked.
- If $\Delta f/2 < B_{\text{PLL}}$, that is, half of the subcarrier spacing smaller than phase noise bandwidth, then the contribution of the ICI on the SNR/EVM is not negligible compared with the CPE because the phase noise from each subcarrier phase noise leaks into the others. Contrary to the CPE, the ICI is much more complex to correct because it is a random process. In the time domain, this corresponds to phase noise variations which are faster than the OFDM symbol time and cannot be tracked.

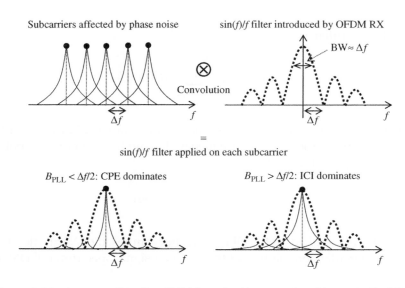

Figure 2.20   In reception the OFDM symbol is convolved by a $\sin(f)/f$ filter

Figure 2.21 shows the phase noise impact on a 64-QAM constellation by considering the same integrated phase noise of 0.056 rad (-25 dBc), but with two different bandwidths. When the phase noise bandwidth is much smaller than the subcarrier spacing, $B_{PLL} = 0.1\Delta f$ (Figure 2.21a), CPE dominates and the common phase rotation is visible on the constellation. On the other hand, when the phase noise bandwidth is much larger than the subcarrier spacing, $B_{PLL} = 10\Delta f$ (Figure 2.21b), ICI dominates and generates Gaussian-type noise. In addition, we observe that the SNR due to phase noise estimated on the constellation, around 25 dB, is in line with Equation 2.43 demonstrated in Section 2.3.4 for integrated phase noise equal to -25 dBc.

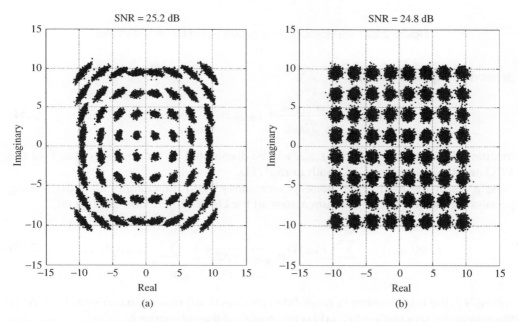

Figure 2.21 Impact of same phase noise power ($\theta_{rms} = 3.22°$ or -25 dBc) on a 64-QAM constellation. (a) $B_{PLL} = 0.1\Delta f$: CPE dominates; (b) $B_{PLL} = 10\Delta f$: ICI dominates

2.3.5.2 Out-of-Band: Reciprocal Mixing due to OFDM Interferers in Reception

As we have seen previously in Section 2.3.1, in reception out-of-band interferers can increase the noise level in the band of interest because of the reciprocal mixing (Figure 2.13). As depicted in Figure 2.22 in the case of an OFDM interferer, due to the convolution in the frequency domain introduced by the mixer (Equation 2.23) the LO phase noise will affect all the interferer subcarriers. If we assume that the frequency offset between the channel and the interferer is much larger than the PLL bandwidth B_{PLL} and the flicker corner frequency, we can approximate the phase noise profile

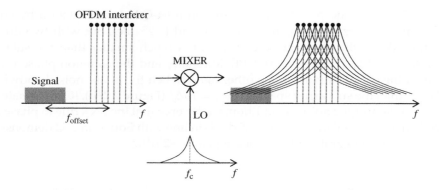

Figure 2.22   Reciprocal mixing due to an OFDM interferer

defined by Equation 2.33 as

$$L(f_{\text{offset}}) = \frac{B_{\text{PLL}}^2 L_0}{f_{\text{offset}}^2} + L_{\text{floor}} = \frac{\alpha}{f_{\text{offset}}^2} + L_{\text{floor}} \tag{2.54}$$

meaning that the phase noise at f_{offset} will be dominated by the component α / f^2 of the VCO and the noise floor L_{floor}, both in rad^2 / Hz.

As a result the total phase noise power aliased into the channel after the mixing process can be derived from a sum across all the OFDM interferer subcarriers:

$$P_{\text{PN}}(f_{\text{offset}}) = \sum_{\substack{k=-N/2 \\ k \neq 0}}^{N/2} P_{\text{sub}}(k) \frac{\alpha}{(f_{\text{offset}} + k\Delta f)^2} + \sum_{\substack{k=-N/2 \\ k \neq 0}}^{N/2} P_{\text{sub}}(k) L_{\text{floor}} \tag{2.55}$$

where N is the total number of modulated carriers ($k \neq 0$ means no carrier at DC), Δf is the subcarrier spacing, and $P_{\text{sub}}(k)$ is the power of the subcarrier k.

If we assume that all the subcarriers have the same power, one can rewrite Equation 2.55 as

$$P_{\text{PN}}(f_{\text{offset}}) = P_{\text{sub}} \left\{ \frac{\alpha}{f_{\text{offset}}^2} \sum_{\substack{k=-N/2 \\ k \neq 0}}^{N/2} \frac{1}{[1 + (k\Delta f / f_{\text{offset}})]^2} + N L_{\text{floor}} \right\} \tag{2.56}$$

It is interesting to note from Equations 2.55 and 2.56 that the OFDM interferer subcarriers close to the channel will introduce more noise in the band of interest due to the fact that the phase noise is inversely proportional to $(f_{\text{offset}} + k\Delta f)^2$. This is the case for an adjacent channel, for example.

But if the frequency offset is much larger than $k\Delta f$, all the subcarriers will have approximately the same contribution and the OFDM interferer can be approximated as a single carrier interferer (having a power NP_{sub}) from a reciprocal mixing point of view:

$$P_{PN}(f_{offset}) \approx NP_{sub}\left(\frac{\alpha}{f_{offset}^2} + L_{floor}\right) \tag{2.57}$$

2.4 Sampling Jitter

In a modern transceiver, the DAC and ADC delimit the boundary between the digital and analog parts of the signal processing. The clocks used to sample the signal are not ideal and introduce error on the sampling time which manifests itself as noise and can limit the system performance. This timing error is known as jitter, and is expressed in seconds.

2.4.1 Jitter Definitions

A real oscillator will always suffer from phase noise (in the frequency domain), which can equivalently be described as jitter in the time domain reflecting random fluctuations in period (Drakhlis, 2001; Mansuri and Yang, 2002). If we define T_0 as the mean period of the oscillator (generally approximated as the ideal one) and its nth period as T_n, the period error $\Delta T_n = (T_n - T_0)$ is called *period jitter* or *cycle jitter* (Silicon Laboratories, 2006) and is generally defined with its STD value:

$$\sigma_c = \sqrt{\frac{1}{N}\sum_{n=1}^{N}\Delta T_n^2} = \sqrt{\frac{1}{N}\sum_{n=1}^{N}(T_n - T_0)^2} \tag{2.58}$$

Cycle jitter compares the oscillating period with the mean period, while *cycle-to-cycle jitter* compares each period with the preceding period. Cycle-to-cycle jitter describes the short-term dynamics of the period; its STD is given by

$$\sigma_{cc} = \sqrt{\frac{1}{N}\sum_{n=1}^{N}(T_{n+1} - T_n)^2} \tag{2.59}$$

Another definition which is very important in communication systems is the *cumulative jitter* (Awad, 1995) because it is function of time as opposed to period and cycle-to-cycle jitters. Cumulative jitter represents the cumulated period errors after a certain number of periods N, its STD is

$$\sigma_{cumul}(N) = \sqrt{\sum_{n=1}^{N}\Delta T_n^2} \tag{2.60}$$

If we assume that the period errors ΔT_n are independent, that is, the sum in Equation 2.60 is incoherent, we can express cumulative jitter STD as a function of the cycle jitter one defined by Equation 2.58:

$$\sigma_{\text{cumul}}(N) = \sqrt{N}\sigma_{\text{c}} \tag{2.61}$$

Figure 2.23 illustrates how cumulative jitter STD cumulates cycle after cycle whereas the cycle/period jitter STD does not depend on the time. For a free-running oscillator cumulative jitter will diverge as time goes to infinity. In a closed-loop system like a PLL the absolute jitter can be bounded.

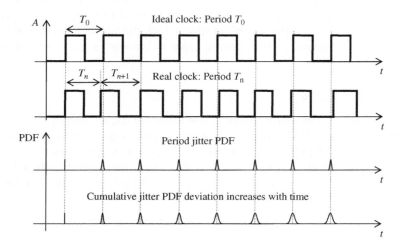

Figure 2.23   Cumulative jitter standard deviation increases at each clock cycle

2.4.2 Sampling Jitter and Phase Noise Relationship

Let us consider a real sampling clock:

$$\text{Clk}(t) = \sin[2\pi F_s t + \theta(t)] = \sin\{2\pi F_s[t + j(t)]\} \tag{2.62}$$

F_s being the sampling frequency in hertz, $\theta(t)$ the phase jitter in radians, and $j(t)$ the timing jitter in seconds.

From Equation 2.62 we can derive the relationship between phase jitter and timing jitter:

$$j(t) = \frac{\theta(t)}{2\pi F_s} \tag{2.63}$$

giving the phase jitter RMS in radians

$$\theta_{\text{RMS}} = \sqrt{\left\langle |\theta(t)|^2 \right\rangle} = \sqrt{4\pi^2 F_s^2 \sigma_j^2} \tag{2.64}$$

where the timing jitter RMS in seconds is defined as

$$\sigma_j = \sqrt{\left\langle \left| j(t) \right|^2 \right\rangle} \tag{2.65}$$

If we suppose that the signal is sampled with a nominal sampling period $T_s = 1/F_s$, due to the timing jitter $j(t)$ the actual sampling instants become

$$t(n) = nT_s + j(nT_s) = nT_s + \frac{\theta(nT_s)}{2\pi F_s} \tag{2.66}$$

In Equation 2.66 the timing jitter becomes the sampling jitter.

For the rest of the study, let us suppose that the instantaneous sampling period is affected by a cycle jitter $j_c(n)$

$$T_s(n) = T_s + j_c(n) \tag{2.67}$$

in order to rewrite Equation 2.66 as a function of the cumulative jitter over n sampling periods:

$$t(n) = nT_s + j(n) = nT_s + \frac{\theta(n)}{2\pi F_s} = nT_s + \sum_{k=1}^{n} j_c(k) \tag{2.68}$$

Equation 2.68 demonstrates that the sampling jitter at the time n is in reality the accumulation of the clock cycle jitter over n periods of the sampling clock:

$$j_{\text{cumul}}(n) = j(n) = \frac{\theta(n)}{2\pi F_s} = \sum_{k=1}^{n} j_c(k) \tag{2.69}$$

From Equation 2.69 it is meaningful to see that the cycle jitter of the sampling clock can be estimated from the difference of two consecutive sampling jitter values:

$$j_c(n) = j(n+1) - j(n) = \frac{\theta(n+1) - \theta(n)}{2\pi F_s} \tag{2.70}$$

giving the cycle jitter RMS value defined by the Equation 2.58:

$$\sigma_c = \sqrt{\frac{1}{N} \sum_{n=1}^{N} \left| j(n+1) - j(n) \right|^2} = \sqrt{\frac{1}{N} \sum_{n=1}^{N} \frac{\left| \theta(n+1) - \theta(n) \right|^2}{4\pi^2 F_s^2}} \tag{2.71}$$

For illustration we generated sampling jitter and cycle jitter time vectors, plotted in Figure 2.24, from the phase noise profile represented in Figure 2.17 by assuming a sampling frequency $F_s = 500\,\text{MHz}$. We can visibly observe low-frequency variations in the sampling jitter which are due to the accumulation of the period errors over many periods comparable to a low-pass filtering with high gain. On the other hand, the

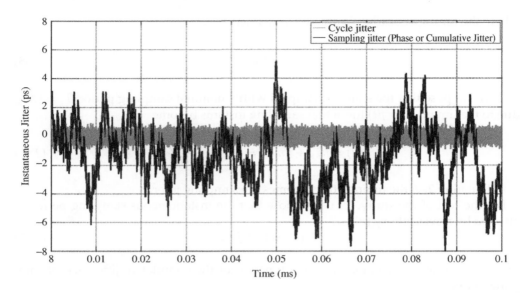

Figure 2.24   Phase jitter and cycle jitter generated from Figure 2.17 phase noise profile

cycle jitter which is estimated as the difference between two consecutive sampling jitter values does not present this low-frequency behavior and exhibits a smaller STD.

Figure 2.25 depicts the PSD of the sampling jitter from $j(n)$ (Equation 2.63) and the PSD of the cycle jitter estimated from $j_c(n)$ (Equation 2.70), but expressed in terms of phase noise (rad^2/Hz in linear, or dBc/Hz in decibels) using these relationships:

$$\theta(n) = 2\pi F_s j(n)$$
$$\theta_c(n) = 2\pi F_s j_c(n) \tag{2.72}$$

As expected, the sampling jitter PSD is equivalent to the oscillator phase noise profile (Figure 2.17) from which it was generated. The low-frequency temporal variations are directly due to the low-frequency oscillator phase noise. On the other hand, we clearly see that the cycle jitter PSD is dominated by the high-frequency phase noise. This is because it has been estimated from the difference between two sampling jitter samples (Equation 2.70). From a mathematical point of view this difference can be seen as a derivative of the sampling jitter by combining Equations 2.70 and 2.72:

$$\theta_c(n) = 2\pi \frac{j(n+1) - j(n)}{T_s} = 2\pi \frac{dj(nT_s)}{T_s} \tag{2.73}$$

Using the fact that the Fourier transform of the signal derivative corresponds to a multiplication in the frequency domain by $j2\pi f$, that is, a first-order high-pass filter (+20 dB/decade):

$$\frac{dg(t)}{dt} \xleftrightarrow{\text{Fourier}} j2\pi f \times G(f) \tag{2.74}$$

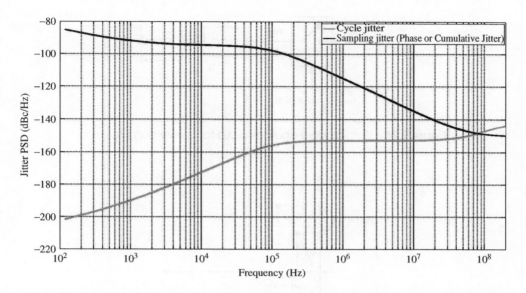

Figure 2.25 PSD estimated from sampling jitter and cycle jitter depicted in Figure 2.24

the PSD of Equation 2.73 becomes

$$\left|FT\left[\theta_c(n)\right]\right|^2 = \left|j4\pi^2 f \times FT\left[j(n)\right]\right|^2 \Rightarrow O_c(f) = 16\pi^4 f^2 O(f) \tag{2.75}$$

where $O_c(f)$ and $O(f)$ are the cycle jitter PSD and the sampling jitter PSD in rad^2/Hz, respectively. Equation 2.75 demonstrates mathematically that the cycle jitter does not present low frequencies due to the multiplication of the phase noise spectrum by f^2. From a physical point of view we can say that the cycle jitter is estimated from two edges separated on average by one clock period, so it makes sense that high-frequency components contribute more than low-frequency ones ($f \ll 1/T_s$).

2.4.3 SNR Limitation due to Sampling Jitter

An ADC is analogous to a switch, which periodically samples an analog signal, followed by a quantizer which produces the digital signal used by the DBB, as presented in Figure 2.26.

If we consider that the input signal $x(t)$ is a sine wave, having a frequency f_0, sampled at a rate of F_s perturbed by a sampling jitter $j(n)$, we obtain

$$x(t = nT_s + j(n)) = A\sin[2\pi f_0(nT_s + j(n))] \tag{2.76}$$

As shown in Figure 2.27, the timing error introduces a sampled amplitude error proportional to the signal derivative:

$$\varepsilon(n) = \frac{\partial x(t)}{\partial t}j(n) = 2\pi f_0 j(n)A\cos(2\pi f_0 t) \tag{2.77}$$

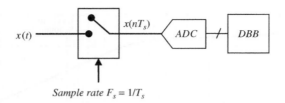

Figure 2.26   Generalized analog-to-digital conversion process

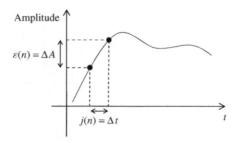

Figure 2.27   Impact of the timing jitter depends on the signal derivative

By assuming that the jitter is a random process uncorrelated with the signal, we can calculate the noise power induced by the clock jitter in the *Nyquist* band $\pm F_s/2$:

$$\langle \varepsilon^2 \rangle = 4\pi^2 f_0^2 A^2 \langle j^2(n) \rangle \frac{1}{T_0} \int_0^{T_0} \cos^2(2\pi f_0 t)\, dt \tag{2.78}$$

where $T_0 = 1/f_0$ and the angle brackets denote the average operator.

By resolving Equation 2.78, we obtain

$$\langle \varepsilon^2 \rangle = 2\pi^2 f_0^2 A^2 \sigma_j^2 \tag{2.79}$$

in which σ_j is the RMS of the sampling jitter $j(n)$ integrated over the whole Nyquist band. The clock-limited SNR as a function of the sampling clock timing jitter is given by

$$\mathrm{SNR_j} = \frac{A^2/2}{\langle \varepsilon^2 \rangle} = \frac{1}{4\pi^2 f_0^2 \sigma_j^2} \tag{2.80}$$

It is interesting to notice that the SNR limitation due to the sampling jitter is inversely proportional to the signal frequency. It means that high-frequency signals are more sensitive to the jitter because their periods tend to reach the jitter STD. From

Equation 2.64, we know that phase jitter and timing jitter are linked:

$$\sigma_j^2 = \frac{\theta_{RMS}^2}{4\pi^2 F_s^2} \tag{2.81}$$

So Equation 2.80 can also be expressed as a function of the integrated phase noise:

$$SNR_j = \frac{1}{\theta_{RMS}^2} \left(\frac{F_s}{f_0} \right)^2 \tag{2.82}$$

In reality, Equations 2.80 and 2.82 express the same thing, but the latter is more convenient because the integrated phase noise is more commonly used in design and can be estimated from the sampling clock phase noise profile.

The RMS phase jitter can also be estimated over a limited bandwidth B by integrating the sampling clock phase noise in the frequency domain:

$$\theta_{RMS}^2 = 4\pi^2 F_s^2 \sigma_j^2 = \int_B L(f)\, df \tag{2.83}$$

in which $L(f)$ is the sampling clock phase noise PSD in rad^2/Hz.

2.4.4 Impact of Sampling Jitter in OFDM

We demonstrated in Section 2.3.5 that the phase noise of the LO used for quadrature mixing affects all the subcarriers in the same way, namely by introducing CPE and ICI, and the overall SNR_{PN} is inversely proportional to the LO integrated phase noise (Equation 2.43).

In the case of sampling jitter, Equation 2.82 shows that the SNR is also limited by the sampling clock integrated phase noise but scaled by the ratio F_s/f_0. As a result, the SNR worsens as the signal frequency f_0 increases.

Because OFDM signals are composed of a sum of multiple sine waves, each subcarrier will be affected differently, with high-frequency subcarriers being more sensitive to the sampling clock jitter, as illustrated in Figure 2.28.

Replacing f_0 by $k\Delta f$ in Equations 2.80 and 2.82, we obtain the SNR per subcarrier:

$$SNR_j(k) = \frac{1}{4\pi^2 (k\Delta f)^2 \sigma_j^2} = \frac{1}{\theta_{RMS}^2} \left(\frac{F_s}{k\Delta f} \right)^2 \tag{2.84}$$

where k is the subcarrier index $[-N/2$ to $N/2 - 1]$ and Δf is the subcarrier spacing.

2.4.5 Sampling Jitter Modeling

In the previous sections we have seen that due to the clock phase jitter $j(t)$ the signal $x(t)$ is sampled at $nT_s + j(nT_s)$ instead of the ideal instants nT_s; consequently, for system

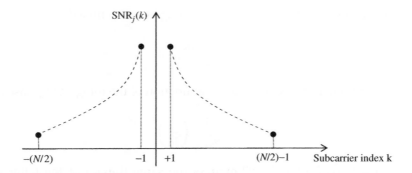

Figure 2.28   SNR limitation per subcarrier due to the sampling jitter

simulations we need to generate a timing jitter $j(nT_s)$ to add to the ideal sampling instants. Because the clock phase jitter is generally specified in the frequency domain using its PSD in rad^2/Hz (or dBc/Hz), we can reuse the IFT method described in Section 2.3.2 so as to generate the timing jitter from the phase noise spectrum $O(f)$ and Equation 2.63 as

$$j(nT_s) = \frac{\theta(nT_s)}{2\pi F_s} = \frac{\text{Re}\{\text{IFT}[O(k\Delta f)]\}}{2\pi F_s} \tag{2.85}$$

where the IFT is performed over N points and $1/NT_s$ is the frequency resolution.

Consequently, the new time vector including the sampling jitter becomes

$$nT_s + j(nT_s) \tag{2.86}$$

For simulating the clock sampling jitter the ideal signal $x(nT_s)$ has to be resampled, or interpolated using the timing vector defined by the Equation 2.86:

$$x(nT_s) \xrightarrow{\text{Resample}} x\left(nT_s + j(nT_s)\right) \tag{2.87}$$

2.5 Carrier Frequency Offset

2.5.1 Description

In communication systems the carrier frequencies are generated from frequency synthesizers, commonly PLLs, using precise crystal oscillators whose performance is specified with a certain accuracy in parts per million (ppm). Because TX and RX carrier frequencies will be slightly different, after frequency down-conversion in reception the signal will have a residual frequency error, which is referred to as CFO.

Figure 2.29 illustrates this impairment. Due to the LO frequency difference between the transmitter and the receiver, after frequency down-conversion to low-IF or zero-IF

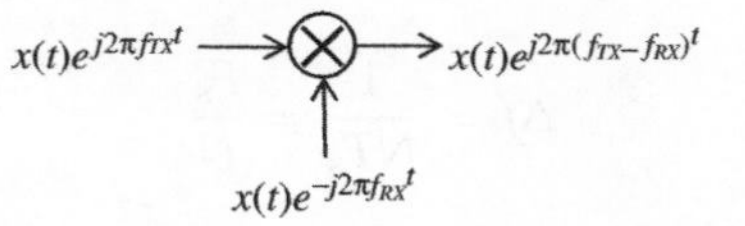

$$x(t)e^{j2\pi f_{TX}t} \longrightarrow \bigotimes \longrightarrow x(t)e^{j2\pi(f_{TX}-f_{RX})t}$$

$$x(t)e^{-j2\pi f_{RX}t}$$

Figure 2.29   Carrier frequency offset after frequency mixing in reception

the signal spectrum will be shifted by a frequency offset:

$$y_{BB}(t) = x(t)\, e^{j2\pi(f_{TX}-f_{RX})t} = x(t)\, e^{j2\pi \Delta_{CFO}t} \tag{2.88}$$

in which Δ_{CFO} is the spectral shift introduced by the CFO.

2.5.2 Impact of CFO in OFDM

By assuming a CFO Δ_{CFO} after frequency down-conversion in the receiver, the OFDM baseband signal can be written

$$y(n) = x(nT_s)\, e^{j2\pi \Delta_{CFO} n T_s} = \left[\sum_{k=-N/2}^{N/2-1} s_k\, e^{j2\pi(kn/N)} \right] e^{j2\pi \Delta_{CFO} n T_s} \tag{2.89}$$

in which N is the total number of subcarriers within the OFDM symbol (including DC and guard bands), s_k is the symbol modulating subcarrier k, and T_s is the sampling frequency. For data demodulation the OFDM receiver will apply a DFT:

$$Y(m) = \frac{1}{N} \sum_{n=0}^{N-1} y(n)\, e^{-j2\pi(mn/N)} = \frac{1}{N} \sum_{k=-N/2}^{N/2-1} s_k \sum_{n=0}^{N-1} e^{-j2\pi\{[(m-k)/N]-\Delta_{CFO}T_s\}n} \tag{2.90}$$

By using the result of the geometric series

$$\sum_{n=0}^{N-1} z^{-n} = \frac{1-z^{-N}}{1-z^{-1}} \tag{2.91}$$

we can develop Equation 2.90:

$$Y(m) = \frac{1}{N} \sum_{k=-N/2}^{N/2-1} s_k \frac{1 - e^{-j2\pi[m-k-(\Delta_{CFO}/\Delta f)]}}{1 - e^{-j(2\pi/N)[m-k-(\Delta_{CFO}/\Delta f)]}}$$

$$= \sum_{k=-N/2}^{N/2-1} s_k \frac{\sin\left[\pi\left(m-k-\dfrac{\Delta_{CFO}}{\Delta f}\right)\right]}{N\sin\left[\dfrac{\pi}{N}\left(m-k-\dfrac{\Delta_{CFO}}{\Delta f}\right)\right]} e^{-j\pi[(N-1)/N][m-k-(\Delta_{CFO}/\Delta f)]} \tag{2.92}$$

with the subcarrier spacing

$$\Delta f = \frac{1}{NT_s} = \frac{F_s}{N} \tag{2.93}$$

The sum in Equation 2.92 can be decomposed into two components to analyze separately:

- Amplitude distortion and phase rotation for $m = k$

$$Y_{m=k}(k) = s_k \frac{\sin\left(\pi \frac{\Delta_{CFO}}{\Delta f}\right)}{N \sin\left(\frac{\pi}{N} \frac{\Delta_{CFO}}{\Delta f}\right)} e^{j\pi[(N-1)/N](\Delta_{CFO}/\Delta f)} \tag{2.94}$$

Equation 2.94 shows that all the subcarriers will be affected by the same phase rotation $\Delta \varphi$ and the same attenuation ΔA depending on the CFO. If $N \gg 1$, one can approximate these two impairments as

$$\Delta A = \frac{\sin\left(\pi \frac{\Delta_{CFO}}{\Delta f}\right)}{N \sin\left(\frac{\pi}{N} \frac{\Delta_{CFO}}{\Delta f}\right)} \tag{2.95}$$

$$\Delta \varphi = \pi \frac{\Delta_{CFO}}{\Delta f}$$

From one symbol to the next, CFO will introduce another deterministic phase increment to the subcarriers depending on the OFDM symbol duration including the guard interval. If we assume the first symbol as the reference, $i = 1$, the phase rotation introduced by CFO for the symbol i is

$$\Delta \varphi(i) = \pi \frac{\Delta_{CFO}}{\Delta f} + 2\pi \Delta_{CFO}(i-1)N_s T_s \tag{2.96}$$

where i is symbol index and $N_s = N_g + N$ is the number of points of a single OFDM symbol including the guard interval.
- ICI for $m \neq k$:

$$Y_{m \neq k}(m) = \sum_{\substack{k=-N/2 \\ m \neq k}}^{N/2-1} s_k \frac{\sin\left[\pi\left(m-k-\frac{\Delta_{CFO}}{\Delta f}\right)\right]}{N \sin\left[\frac{\pi}{N}\left(m-k-\frac{\Delta_{CFO}}{\Delta f}\right)\right]} e^{-j\pi[(N-1)/N][m-k-(\Delta_{CFO}/\Delta f)]} \tag{2.97}$$

Equation 2.97 shows the ICI. Indeed, in addition to amplitude distortion and phase rotation, all the subcarriers will be distorted by the $(N-1)$ others. For small Δ_{CFO}

the normalized ICI noise power introduced by CFO can be approximated as the following (Pollet *et al.*, 1995):

$$\sigma^2_{\text{ICI\_CFO}} = \frac{\pi^2}{3} \left(\frac{\Delta_{\text{CFO}}}{\Delta f} \right)^2 \tag{2.98}$$

Figure 2.30 schematizes the impact of the CFO due to the frequency shift of the receiver FFT weighting function "sin$(f)/f$" breaking the orthogonality between the subcarriers. We can clearly see the common amplitude attenuation and phase shift for the considered subcarriers k, and the ICI created by the other subcarriers ($\ldots k-2$, $k-1$, $k+1$, $k+2$, etc.) which are not null anymore because they are no longer aligned with the zeros of the FFT weighting function.

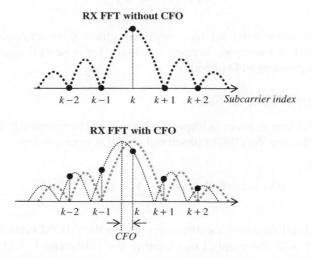

Figure 2.30   Illustration of the CFO impact after the FFT

2.5.2.1 CFO Modeling

It is very simple to model the CFO by merely frequency shifting the baseband signal as written in Equation 2.88.

2.6 Sampling Frequency Offset

2.6.1 Description

Like the carrier frequencies, transmitter and receiver sampling frequencies used by the DAC and ADC are generally slightly mismatched. This impairment is known as SFO and will degrade the system performance.

Due to the SFO the received signal $x(t)$ is sampled at $n(T_s + \Delta T)$ instead of the ideal nT_s:

$$x(nT_s) \xrightarrow{\text{SFO}} x(n(T_s + \Delta T)) \tag{2.99}$$

The SFO is defined as

$$\text{SFO} = \frac{1}{T_s} - \frac{1}{T_s + \Delta T} = F_s \frac{\Delta T / T_s}{1 + (\Delta T / T_s)} \tag{2.100}$$

in which F_s is the ideal sampling frequency, $F_s = 1/T_s$, and $\Delta T/T_s$ is the relative sampling clock offset. Because $\Delta T/T_s$ is generally specified in ppm and then much smaller than one, we can approximate Equation 2.100 as

$$\text{SFO} \approx F_s \frac{\Delta T}{T_s} = F_s \delta \tag{2.101}$$

in which δ is the normalized relative sampling clock accuracy/precision generally specified in ppm. For example, 50 ppm precision for a sampling clock running at 100 MHz means a possible SFO of 5 kHz.

2.6.2 Impact of SFO in OFDM

Let us suppose that the receiver is impaired by an SFO but perfectly synchronized on the symbols; in this case the OFDM baseband signal is expressed as

$$y(n) = x(nT_s(1 + \delta)) = \sum_{k=-N/2}^{N/2-1} s_k \, e^{j(2\pi/N)k(1+\delta)n} \tag{2.102}$$

in which N is the total number of subcarriers within the OFDM symbol (including DC and guard bands), s_k is the symbol modulating the subcarrier k, and δ is the relative SFO (normalized by the ideal sampling frequency).

The OFDM receiver will apply a DFT:

$$Y(m) = \frac{1}{N} \sum_{n=0}^{N-1} y(n) \, e^{-j2\pi(mn/N)} = \frac{1}{N} \sum_{k=-N/2}^{N/2-1} s_k \sum_{n=0}^{N-1} e^{-j(2\pi/N)[m-k(1+\delta)]n} \tag{2.103}$$

By the use of Equation 2.91, we can develop Equation 2.103:

$$Y(m) = \sum_{k=-N/2}^{N/2-1} s_k \frac{\sin\{\pi[m - k(1 + \delta)]\}}{N \sin\left\{\frac{\pi}{N}[m - k(1 + \delta)]\right\}} e^{-j\pi[(N-1)/N][m-k(1+\delta)]} \tag{2.104}$$

By comparing Equations 2.92 and 2.104, we see that the impact of the SFO is quite similar to CFO and can be decomposed into two components to analyze separately:

- Amplitude distortion and phase rotation for $m = k$

$$Y_{m=k}(k) = s_k \frac{\sin(\pi k \delta)}{N \sin\left(\frac{\pi}{N} k \delta\right)} \, e^{j\pi[(N-1)/N]k\delta} \tag{2.105}$$

Equation 2.105 shows that all the subcarriers are affected by a phase shift and an amplitude distortion similar to CFO, but in the case of SFO both the effects depend on the subcarrier index k. If $N \gg 1$:

$$\Delta A(k) = \frac{\sin(\pi k \delta)}{N \sin\left(\frac{\pi}{N} k \delta\right)} \tag{2.106}$$

$$\Delta \varphi(k) = k\pi \delta$$

The phase shift is linearly increasing with a difference of $\pi \delta$ between two consecutive subcarriers; this property will be useful for the SFO estimation and correction algorithms.

- ICI for $m \neq k$:

$$Y_{m \neq k}(m) = \sum_{\substack{k=-N/2 \\ m \neq k}}^{N/2-1} s_k \frac{\sin\{\pi[m - k(1+\delta)]\}}{N \sin\left\{\frac{\pi}{N}[m - k(1+\delta)]\right\}} \, e^{-j\pi[(N-1)/N][m-k(1+\delta)]} \tag{2.107}$$

As with the CFO, this term represents ICI, that is, a distortion caused by the $(N-1)$ other subcarriers affecting the subcarrier under consideration. For small δ the normalized ICI noise power introduced by SFO can be approximated to (Pollet et al., 1994)

$$\sigma_{\text{ICI\_SFO}}^2(k) = \frac{\pi^2}{3}(k\delta)^2 \tag{2.108}$$

We can note that the ICI noise introduced by SFO depends on the subcarrier index and augment as the subcarrier frequency increases, as opposed to the ICI noise generated by CFO which is constant (Equation 2.98).

Figure 2.31 presents the effect of the SFO on one subcarrier. The zeros of the FFT weighting function are shifted compared with the original subcarrier positions k because they are now fixed by the new indices m shifted by the SFO. As a result this sampling frequency shift introduces amplitude attenuation and phase shift of the considered subcarriers k, and ICI because the non-observed subcarriers are not null anymore. On the other hand, as opposed to the CFO, the SFO impact depends on the subcarrier index k.

The SFO will also impact the OFDM symbol time synchronization which consists of finding the precise instant when the symbol starts before performing the FFT. It is illustrated in Figure 2.32. If we assume that the time synchronization is perfectly adjusted

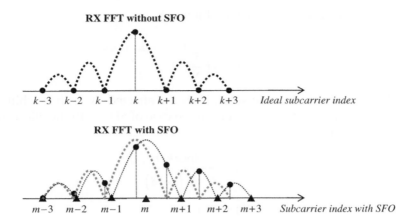

Figure 2.31   Illustration of the SFO after the FFT

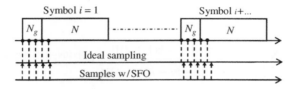

Figure 2.32   OFDM symbol de-synchronization due to SFO

on the first OFDM symbol, the SFO will create a drift of the time synchronization for the following symbols:

$$\Delta t(i) = (i - 1)N_s T_s \delta \tag{2.109}$$

where i is symbol index and N_s is the number of points of a single OFDM symbol including the guard interval ($N_s = N_g + N$).

The consequence of this OFDM symbol time synchronization error is another phase rotation linearly increasing with the symbol indices. As a result the phase shift present in the Equation 2.106 becomes

$$\Delta\varphi(k, i) = k\pi\delta + 2\pi k\frac{F_s}{N}\Delta t(i) = k\pi\delta + \frac{2\pi}{N}k(i - 1)N_s\delta \tag{2.110}$$

2.6.2.1 SFO Modeling

SFO modeling is more complex than CFO modeling due to the need to interpolate the signal on a new time vector, compressed or expanded. If we consider F_s as the ideal sampling frequency and δ its accuracy, we need to create the time vector $nT_s(1+\delta)$ and to resample the signal at the new time instants:

$$x(nT_s) \xrightarrow{\text{Resample}} x(nT_s(1+\delta)) \tag{2.111}$$

2.7 I and Q Mismatch

2.7.1 Description

As we have seen earlier, mixers are used in communication systems for up-converting the signal to RF for transmission, and for down-converting to baseband for reception. But for fully utilizing the bandwidth of the signal it is preferable to use quadrature mixers, also called complex mixers because they operate on the signal in the complex domain. Indeed, by using the real and imaginary parts of the signal, quadrature mixers provide the possibility to distinguish the "positive and negative" frequencies for a complex baseband spectrum centered around DC. In real implementations quadrature mixers are impaired by gain and phase mismatches which affect the signal by degrading the SNR or by generating out-of-band noise.

The normalized gain imbalance ΔG quantifies the difference between the gains G_I in the real branch (I) and G_Q in the imaginary branch (Q):

$$\Delta G = \frac{|G_I - G_Q|}{G_Q} \tag{2.112}$$

which is often specified as a percentage. However, sometimes the gain imbalance is specified in decibels:

$$\Delta G_{dB} = 20 \log_{10}\left(\frac{G_I}{G_Q}\right) > 0\,\text{dB} \tag{2.113}$$

and the relationship between Equations 2.112 and 2.113 is given by

$$\Delta G = 10^{\Delta G_{dB}/20} - 1 \tag{2.114}$$

The phase imbalance $\Delta\phi$ is the phase error between the components I and Q ideally separated by 90°:

$$\text{phase(I)} - \text{phase(Q)} = \pm\frac{\pi}{2} + \Delta\phi \tag{2.115}$$

Transmitter and receiver IQ mismatches are separately represented in Figure 2.33. We will use ideal receiver and ideal transmitter models for separating the IQ mismatches in transmission and reception, respectively.

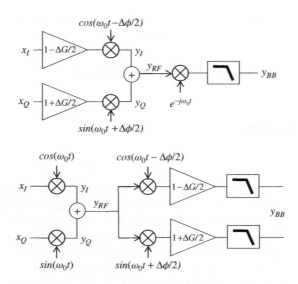

Figure 2.33 IQ mismatch models for transmission and reception

2.7.1.1 Transmitter IQ Mismatch

Let us write the transmitter RF signal taking into account the quadrature mismatch:

$$y_{RF}(t) = \left(1 - \frac{\Delta G}{2}\right) x_I \cos(\omega_0 t - \Delta\phi/2) + \left(1 + \frac{\Delta G}{2}\right) x_Q \sin(\omega_0 t + \Delta\phi/2) \qquad (2.116)$$

in which x_I and x_Q are the real and imaginary components of the complex baseband signal, respectively.

In order to only study the transmitter quadrature mismatch we assume that the RF signal is frequency down-converted by an ideal receiver (top of Figure 2.33):

$$y_{BB}(t) = y_{RF}(t)\, e^{-j\omega_0 t} = y_{RF}(t)[\cos(\omega_0 t) - j\sin(\omega_0 t)] \qquad (2.117)$$

After some mathematical manipulations and by supposing that the high-frequency components around $2\omega_0$ are removed by the receive low-pass filtering, we obtain

$$\begin{aligned}
y_{BB}(t) = {} & \frac{1 - (\Delta G/2)}{2}[\cos(\Delta\phi/2) + j\sin(\Delta\phi/2)]x_I \\
& + \frac{1 + (\Delta G/2)}{2}[\sin(\Delta\phi/2) + j\cos(\Delta\phi/2)]x_Q
\end{aligned} \qquad (2.118)$$

which can be rearranged as follows:

$$y_{BB}(t) = \alpha_{TX}(x_I + jx_Q) + \beta_{TX}(x_I - jx_Q)$$

$$\begin{cases} \alpha_{TX} = \dfrac{1}{2}\left[\cos(\Delta\phi/2) - j\dfrac{\Delta G}{2}\sin(\Delta\phi/2)\right] \\[3mm] \beta_{TX} = \dfrac{1}{2}\left[-\dfrac{\Delta G}{2}\cos(\Delta\phi/2) + j\sin(\Delta\phi/2)\right] \end{cases} \tag{2.119}$$

It is interesting to notice from Equation 2.119 that IQ mismatch generates an image which is the complex conjugate of the original signal. The ratio between the image and the signal is called the IQ mismatch image suppression ratio (ISR):

$$\mathrm{ISR_{TX}} = \frac{|\beta_{TX}|^2}{|\alpha_{TX}|^2} = \frac{\dfrac{\Delta G^2}{4}\cos^2(\Delta\phi/2) + \sin^2(\Delta\phi/2)}{\cos^2(\Delta\phi/2) + \dfrac{\Delta G^2}{4}\sin^2(\Delta\phi/2)} = \frac{\dfrac{\Delta G^2}{4} + \tan^2(\Delta\phi/2)}{1 + \dfrac{\Delta G^2}{4}\tan^2(\Delta\phi/2)} \tag{2.120}$$

By assuming that ΔG and $\Delta\phi$ are normally much smaller than one, the ISR can be approximated as

$$\mathrm{ISR_{TX}} = \frac{\Delta G^2 + \Delta\phi^2}{4} \tag{2.121}$$

In order to illustrate the transmitter IQ mismatch, we will use as baseband input a single tone having frequency Δf:

$$x_I + jx_Q = \cos(2\pi\,\Delta ft) + j\sin(2\pi\,\Delta ft) = e^{j2\pi\,\Delta ft} \tag{2.122}$$

After up-conversion around f_0 by the impaired mixer, the tone frequency is shifted to $(f_0 + \Delta f)$, but due to the IQ mismatch another tone appears at $(f_0 - \Delta f)$:

$$y_{RF} = \alpha_{TX}\,e^{j2\pi(f_0+\Delta f)t} + \beta_{TX}\,e^{j2\pi(f_0-\Delta f)t} = (\alpha_{TX}\,e^{j2\pi\,\Delta ft} + \beta_{TX}\,e^{-j2\pi\,\Delta ft})\,e^{j2\pi f_0 t} \tag{2.123}$$

Figure 2.34 presents the output spectrum of the quadrature mixer after the up-conversion of the single tone. As revealed by Equation 2.123, we retrieve the tone at $(f_0 + \Delta f)$ and the IQ mismatch image at $(f_0 - \Delta f)$ scaled by the ISR.

Because the LO leakage, which is another well-known impairment of up-conversion mixers, can be simply modeled as a sum, as we will see later, we also represented it on the figure.

The effects of the TX IQ mismatch and the LO leakage in zero-IF and low-IF architectures are presented in Figure 2.35.

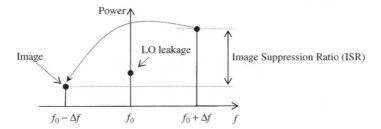

Figure 2.34 Illustration of the LO leakage and the IQ mismatch image after the frequency up-conversion of a single tone

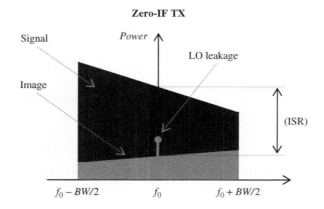

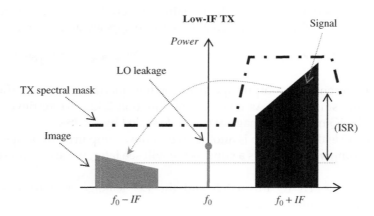

Figure 2.35 Impact of TX IQ mismatch in zero- and low-IF architectures

For zero-IF architectures the complex image of the signal falls exactly within the channel bandwidth, creating a co-channel interference limiting the EVM due to the quadrature mixer ISR.

In low-IF architectures it is not the EVM requirement but the spectral mask, out-of-band emissions, or adjacent channel leakage ratio (ACLR) specifications which will dictate the quadrature mixer performance.

2.7.1.2 Receiver IQ Mismatch

Let us write the received baseband signal assuming an ideal transmitter and taking into account the receiver quadrature mismatch (bottom of Figure 2.33):

$$y_{BB}(t) = [x_I \cos(\omega_0 t) + x_Q \sin(\omega_0 t)]$$

$$\times \left[\left(1 - \frac{\Delta G}{2}\right) \cos(\omega_0 t - \Delta\phi/2) + j \left(1 + \frac{\Delta G}{2}\right) \sin(\omega_0 t + \Delta\phi/2) \right] \quad (2.124)$$

By supposing that the high-frequency components around $2\omega_0$ are removed by the receive low-pass filtering, we obtain

$$y_{BB}(t) = \frac{1}{2} \left[\left(1 - \frac{\Delta G}{2}\right) \cos(\Delta\phi/2) + j \left(1 + \frac{\Delta G}{2}\right) \sin(\Delta\phi/2) \right] x_I$$

$$+ \frac{1}{2} \left[\left(1 - \frac{\Delta G}{2}\right) \sin(\Delta\phi/2) + j \left(1 + \frac{\Delta G}{2}\right) \cos(\Delta\phi/2) \right] x_Q \quad (2.125)$$

As for the transmitter we can rearrange Equation 2.125 in order to show that the receiver IQ mismatch generates an image which is the complex conjugate of the signal:

$$y_{BB}(t) = \alpha_{RX}(x_I + jx_Q) + \beta_{RX}(x_I - jx_Q)$$

$$\begin{cases} \alpha_{RX} = \frac{1}{2} \left[\cos(\Delta\phi/2) + j\frac{\Delta G}{2} \sin(\Delta\phi/2) \right] \\ \beta_{RX} = \frac{1}{2} \left[-\frac{\Delta G}{2} \cos(\Delta\phi/2) + j \sin(\Delta\phi/2) \right] \end{cases} \quad (2.126)$$

The impact of I and Q mismatch in reception is exactly equivalent to that observed in transmission, in that an image is generated which is the complex conjugate of the signal. As a result the receiver ISR derived from Equation 2.126 follows the transmitter case (Equation 2.120) and can also be approximated if ΔG and $\Delta\phi$ are assumed much smaller than one:

$$\text{ISR}_{RX} = \frac{\Delta G^2 + \Delta\phi^2}{4} \quad (2.127)$$

The main difference with transmission IQ mismatch concerns the impact on low-IF architectures. In low-IF, the reception IQ mismatch can result in the aliasing of an interferer or an adjacent channel into the channel bandwidth, thus limiting the receiver SNR. This effect is presented in Figure 2.36.

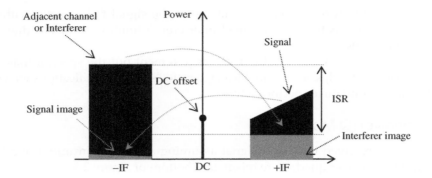

Figure 2.36   Impact of RX IQ mismatch in low-IF architecture

2.7.2 IQ Mismatch Modeling

In order to simplify the notation for the modeling, we will use a matrix form:

$$y_{BB}(t) = \begin{bmatrix} 1 & j \end{bmatrix} \times \mathbf{G_M} \times \mathbf{x} \qquad (2.128)$$

$\mathbf{G_M}$ being the IQ mismatch matrix and $\mathbf{x}$ the baseband input vector. For the transmission, from Equation 2.118 we can rewrite the result of the TX IQ mismatch as

$$y_{BB}(t) = \begin{bmatrix} 1 & j \end{bmatrix} \times \begin{bmatrix} (1 - \Delta G/2)\cos(\Delta\phi/2) & (1 + \Delta G/2)\sin(\Delta\phi/2) \\ (1 - \Delta G/2)\sin(\Delta\phi/2) & (1 + \Delta G/2)\cos(\Delta\phi/2) \end{bmatrix} \times \begin{bmatrix} x_I(t) \\ x_Q(t) \end{bmatrix} \qquad (2.129)$$

In this model it is also possible to introduce a DC component at the input of the mixer for modeling the LO leakage created by the baseband DC offsets:

$$y_{BB}(t) = \begin{bmatrix} 1 & j \end{bmatrix} \times \mathbf{G_M} \times (\mathbf{x} + \mathbf{dc}) \qquad (2.130)$$

$\mathbf{dc} = \begin{bmatrix} dc_I \\ dc_Q \end{bmatrix}$ being the complex DC offset vector.

Regarding reception, from Equation 2.125 we can rewrite the result of the RX IQ mismatch as

$$y_{BB}(t) = \begin{bmatrix} 1 & j \end{bmatrix} \times \begin{bmatrix} (1 - \Delta G/2)\cos(\Delta\phi/2) & (1 - \Delta G/2)\sin(\Delta\phi/2) \\ (1 + \Delta G/2)\sin(\Delta\phi/2) & (1 + \Delta G/2)\cos(\Delta\phi/2) \end{bmatrix} \times \begin{bmatrix} x_I(t) \\ x_Q(t) \end{bmatrix} \qquad (2.131)$$

2.7.3 SNR Limitation due to IQ Mismatch

Let us suppose a transmitted noiseless symbol

$$s = x_I + jx_Q = \cos(\theta) + j\sin(\theta) \qquad (2.132)$$

By introducing TX IQ mismatch and by assuming an ideal receiver, using Equation 2.129 we can express the impaired symbol as

$$
s_{IQ} = \begin{bmatrix} 1 & j \end{bmatrix} \times \begin{bmatrix} (1 - \Delta G/2)\cos(\Delta\phi/2) & (1 + \Delta G/2)\sin(\Delta\phi/2) \\ (1 - \Delta G/2)\sin(\Delta\phi/2) & (1 + \Delta G/2)\cos(\Delta\phi/2) \end{bmatrix} \times \begin{bmatrix} \cos(\theta) \\ \sin(\theta) \end{bmatrix} \quad (2.133)
$$

giving an EVM

$$
\mathrm{EVM}_{IQ}^2 = \frac{\left[\left(1 - \dfrac{\Delta G}{2}\right)\cos(\Delta\phi/2)\cos(\theta) + \left(1 + \dfrac{\Delta G}{2}\right)\sin(\Delta\phi/2)\sin(\theta) - \cos(\theta) \right]^2}{\cos^2(\theta) + \sin^2(\theta)}
$$

$$
+ \frac{\left[\left(1 - \dfrac{\Delta G}{2}\right)\sin(\Delta\phi/2)\cos(\theta) + \left(1 + \dfrac{\Delta G}{2}\right)\cos(\Delta\phi/2)\sin(\theta) - \sin(\theta) \right]^2}{\cos^2(\theta) + \sin^2(\theta)} \quad (2.134)
$$

By supposing that ΔG and $\Delta\phi \ll 1$, $\cos(x) \approx 1$ and $\sin(x) \approx x$ for $x \ll 1$, we can approximate Equation 2.134:

$$
\mathrm{EVM}_{IQ}^2 \approx \left(-\frac{\Delta G}{2}\cos(\theta) + \frac{\Delta\phi}{2}\sin(\theta) \right)^2 + \left(\frac{\Delta\phi}{2}\cos(\theta) + \frac{\Delta G}{2}\sin(\theta) \right)^2 \quad (2.135)
$$

Consequently, the EVM introduced by IQ mismatch can be estimated as follows:

$$
\mathrm{EVM}_{IQ}^2 = \frac{\Delta G^2 + \Delta\phi^2}{4} \quad (2.136)
$$

which is again the IQ mismatch ISR defined in Equations 2.121 and 2.127:

$$
\mathrm{EVM}_{IQ}^2 \approx \mathrm{ISR}_{IQ} \quad (2.137)
$$

and thus

$$
\mathrm{SNR}_{IQ} = \frac{1}{\mathrm{EVM}_{IQ}^2} = \frac{1}{\mathrm{ISR}_{IQ}} = \frac{4}{\Delta G^2 + \Delta\phi^2} \quad (2.138)
$$

Figure 2.37 illustrates the effect of the gain and phase mismatch on a 64-QAM constellation. On the left-hand side we isolated the impact of the gain imbalance whereas the constellation on the right-hand side shows the effect of both. In addition we can see that the SNR estimations are quite in line with the theoretical formula demonstrated above (Equation 2.138); indeed, the SNR measured on the constellation with 10% gain imbalance is 26.5 dB, compared with 26 dB in theory; the SNR measured on the constellation with 10% and 3° gain and phase imbalance, respectively, is 24.3 vs. 25 dB in theory.

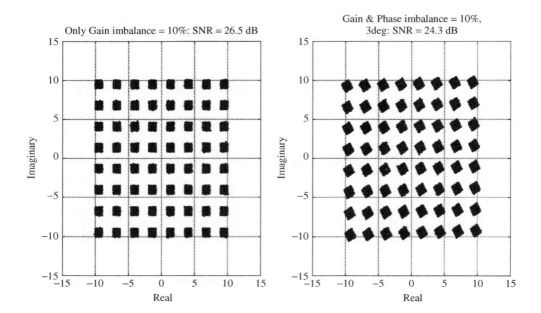

Figure 2.37 Impact of I and Q mismatch on a 64-QAM constellation

2.7.4 Impact of IQ Mismatch in OFDM

If we suppose an ideal OFDM baseband signal $x(t)$, IQ mismatch will generate a complex conjugate image transforming the baseband signal into

$$y_{BB}(n) = \alpha x(n) + \beta x^*(n) = \alpha \sum_{k=-N/2}^{N/2-1} s_k \, e^{j2\pi(kn/N)} + \beta \sum_{k=-N/2}^{N/2-1} s_k^* \, e^{-j2\pi(kn/N)} \qquad (2.139)$$

in which N is the total number of subcarriers within the OFDM symbol (including DC and guard bands), s_k is the data symbol modulating the subcarrier k, and the ratio β/α is the ISR of the quadrature mixer defined by Equation 2.120.

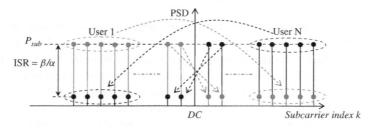

Figure 2.38 Inter-carrier and inter-channel interferences created by IQ mismatch in zero-IF OFDM(A) transceivers

From Equation 2.139 we can see that the original subcarrier k has a mirrored image $-k$ whose amplitude is directly fixed by ISR. For zero-IF OFDM transceivers the impact is ICI because the subcarrier k leaks into the subcarrier $-k$ and vice versa. As a result the mixer ISR becomes a performance-limiting factor (Krone and Fettweis, 2008).

In OFDMA, IQ mismatch also creates inter-channel interference, as illustrated in Figure 2.38.

2.8 DAC/ADC Quantization Noise and Clipping

"Digital to analog" and "analog to digital" converters are the two mixed signal blocks in the transmission and reception paths, respectively, bridging the RF AFE and the DBB. Figure 2.39 presents the transfer function of a 3-bit ADC. We can easily imagine the complementary transfer function for a DAC. For communication systems it is preferable to have a large number of bits in order to reduce the quantization noise and a high clipping level for minimizing the signal saturation. These two impairments will degrade the overall transceiver performance and have to be taken into account in the early stage of system design.

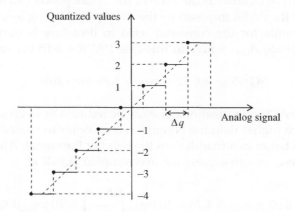

Figure 2.39   Transfer function of a 3-bit ADC

2.8.1 SNR Limitation due to the Quantization Noise and Clipping Level

Quantization noise is generated by the rounding operation performed by the converter when selecting a quantized word from a continuous analog signal. If we define A_{clip} as the converter clipping level and n_{bits} the converter resolution in number of bits, the quantization resolution Δq is given by

$$\Delta q = \frac{2A_{\text{clip}}}{2^{n_{\text{bits}}} - 1} \approx \frac{A_{\text{clip}}}{2^{n_{\text{bits}}-1}} \tag{2.140}$$

if the number $2^{n_{\text{bits}}} \gg 1$. The quantization resolution is in same units as the analog signal.

The average quantization noise power can be computed by assuming the quantization noise distribution is uniform between $-\Delta q/2$ and $+\Delta q/2$:

$$\sigma_N^2 = \int_{-\Delta q/2}^{\Delta q/2} \frac{1}{\Delta q} x^2 \, dx = \frac{1}{\Delta q} \left[\frac{x^3}{3} \right]_{-\Delta q/2}^{\Delta q/2} = \frac{\Delta q^2}{12} \tag{2.141}$$

The converter signal to quantization noise ratio (SQNR) is then defined as the ratio of the signal average power P_s to the quantization noise average power:

$$\text{SQNR} = \frac{P_s}{\sigma_N^2} = \frac{12 P_s}{\Delta q^2} \approx 3 \times \frac{2^{2 n_{\text{bits}}} P_s}{A_{\text{clip}}^2} \tag{2.142}$$

or in decibels:

$$\text{SQNR}_{\text{dB}} \approx 6.02 \times n_{\text{bits}} + 4.76 - 10 \log_{10} \left(\frac{A_{\text{clip}}^2}{P_s} \right) \tag{2.143}$$

in which A_{clip}^2 / P_s can be defined as the ratio of the "peak power" to the average signal power, or in reality the PAPR imposed by the converter clipping level.

The classical formula for the converter used in literature is derived from a sine wave having amplitude A_{clip}, where in this case PAPR $= 3$ dB because $A_{\text{clip}}^2 / P_s = 2$ in linear terms

$$\text{SQNR}_{\text{dB}} \approx 6.02 \times n_{\text{bits}} + 1.76 \text{ for a sine} \tag{2.144}$$

The impact of the ADC quantization noise can be reduced by oversampling the signal at a frequency much higher than the Nyquist rate in order to spread the quantization noise power over a larger bandwidth than the band of the signal. After filtering to the bandwidth of interest, we can express the oversampled SQNR as

$$\text{SQNR}_{\text{dB}} \approx 6.02 \times n_{\text{bits}} + 4.76 - 10 \log_{10} \left(\frac{A_{\text{clip}}^2}{P_s} \right) + 10 \log_{10}(\text{OSR}) \tag{2.145}$$

where OSR is the oversampling ratio defined by the sampling frequency F_s and the signal bandwidth BW.

In some applications, especially those employing wide bandwidths, the performances of the ADC/DAC are not fixed by the quantization noise but by their thermal noise arising from analog circuitry (like the sample-and-hold block for an ADC).

2.8.1.1 ADC Noise Floor versus Receiver Gain

The ADC noise floor will contribute to the overall receiver noise floor fixing the minimum sensitivity level. In order to limit its impact on the system performance, the maximum gain of the receiver can be adjusted in order to neglect its contribution.

Figure 2.40 shows how the gain of the receiver can be adjusted in order to limit the impact of the ADC noise floor on the sensitivity by specifying a noise margin. The main idea is to apply enough gain to amplify the input thermal noise to a level which makes the ADC noise floor negligible by comparison. The receiver sensitivity degradation versus noise margin is plotted in Figure 2.41 in which we can see that an ADC noise

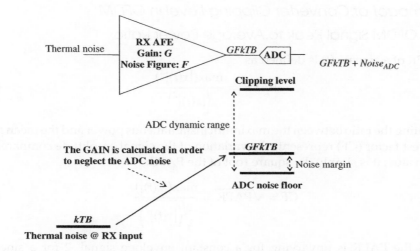

Figure 2.40 Receiver gain specification versus ADC noise contribution

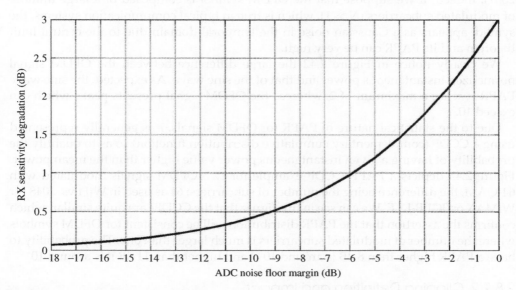

Figure 2.41 RX sensitivity degradation as a function of the ADC noise floor versus AFE noise floor (kTB)

floor 6 dB below the overall receiver noise will degrade the sensitivity by 1 dB. By adding 6 dB more to the margin in order to push the ADC noise floor 12 dB below the receiver noise, that is, increasing the ADC size by one bit, the sensitivity will only be improved by roughly 0.7 dB, which is minimal compared with the design effort to add one bit to the ADC.

2.8.2 Impact of Converter Clipping Level in OFDM

2.8.2.1 OFDM Signal Peak to Average Power Ratio

The PAPR of a signal $x(t)$ is defined as

$$\text{PAPR} = \frac{\max(|x(t)|^2)}{\left\langle |x(t)|^2 \right\rangle} \tag{2.146}$$

representing the ratio between the maximum instantaneous power and the mean power.

The crest factor (CF) represents the variation of the signal amplitude compared with its RMS value; it is simply the square root of the PAPR:

$$\text{CF} = \sqrt{\text{PAPR}} = \frac{\max(|x(t)|)}{\sqrt{\left\langle |x(t)|^2 \right\rangle}} \tag{2.147}$$

Whereas the PAPR is unvarying for a constant envelope signal, 2 for a sine wave (i.e., 3 dB), it can be defined only statistically for an OFDM signal (Ochiai and Imai, 2001). Indeed, if we suppose that the OFDM symbol is composed of a large number of modulated subcarriers ($N \gg 1$), which is true in typical communication systems, the symbol appears as a Gaussian noise in the temporal domain due to the central limit theorem and its PAPR can be very high.

We clearly notice in Figure 2.42 the large difference between the OFDM signal normalized instantaneous power and that of the sine wave. As expected, the sine wave PAPR reaches a maximum of 2, whereas the OFDM signal presents peaks which can exceed 10.

Due to the statistical nature of PAPR for OFDM signals, it is generally represented using a CCDF (complementary cumulative distribution function) so as to quantify the probability of having a given instantaneous power value higher than the mean power. Figure 2.43 depicts a PAPR CCDF example for two OFDM signals modulated with 64-QAM, the difference being the number of subcarriers: 64 as used in WiFi vs. 2048 for WiMAX or 3GPP-LTE. We can see in this figure that the CCDFs are quite similar, which confirms the assertion that the PAPR distribution will be consistent for OFDM symbols where the number of modulated subcarriers is much larger than one. The probability to have a PAPR higher than 8 dB is around 10^{-3} and higher than 11 dB it is around 10^{-6}.

2.8.2.2 Clipping Definition and Impact

The clipping will "cut" the signal and thus limit the signal PAPR, resulting in a degradation of the SNR. Because the OFDM signal is generally complex, clipping

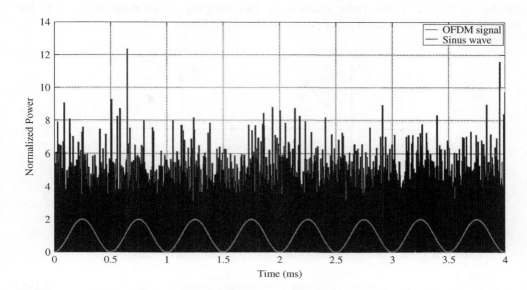

Figure 2.42 Normalized instantaneous power of a sine wave and an OFDM signal

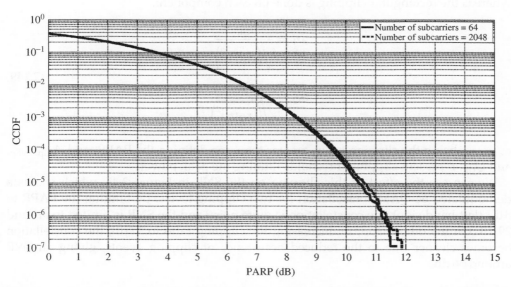

Figure 2.43 Peak to average power ratio CCDF of two OFDM signals: 64 and 2048 subcarriers

can be performed in the polar domain on the complex signal magnitude or in the rectangular domain on each component, as shown in Figure 2.44.

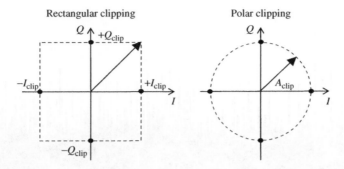

Figure 2.44 Polar and rectangular clippings in the complex plane

The polar clipping is done on the magnitude of the complex signal:

$$\alpha = \frac{A_{\text{clip}}}{\sqrt{\langle |x(t)|^2 \rangle}} = \frac{A_{\text{clip}}}{\sqrt{\langle |x_I(t)^2 + x_Q(t)^2| \rangle}} \tag{2.148}$$

whereas the rectangular clipping is done on each component:

$$\alpha_I = \frac{I_{\text{clip}}}{\sqrt{\langle |x_I(t)|^2 \rangle}}$$

$$\alpha_Q = \frac{Q_{\text{clip}}}{\sqrt{\langle |x_Q(t)|^2 \rangle}} \tag{2.149}$$

α, α_I, α_Q being the clipping ratios greater than one defining the PAPR, and A_{clip}, I_{clip}, and Q_{clip} represent clipping thresholds limiting the peak power.

We can see the effect of this clipping on the SNR in Figure 2.45, which shows a demodulated 64-QAM constellation for two clipping levels. By limiting the PAPR to 8 dB (left-hand side), the constellation is quite noisy and the SNR is limited to around 29 dB. On the other hand, by increasing the clipping ratio to 12 dB, the constellation becomes cleaner due to the reduced occurrence of clipping events, giving an SNR around 56 dB.

2.8.3 DAC and ADC Dynamic Range in OFDM

In order to minimize the size and therefore the cost of on-chip transceivers, the DAC and ADC have to be carefully specified for optimizing the number of bits; that is, their dynamic range. In addition to the required SNDR for the data demodulation, the

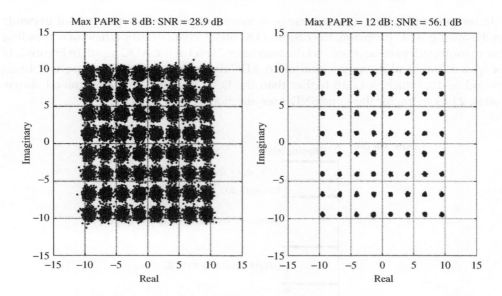

Figure 2.45 Effect of the clipping level on a 64-QAM constellation

system designer must also take into account others factors for calculating the necessary dynamic range.

In transmission the DAC dynamic range includes a noise margin for limiting the DAC noise floor impact on the system, the required SNDR depending on the modulation used, the PAPR of the signal (which is quite large for OFDM signals), and a possible digital gain control range. Figure 2.46 details the TX dynamic range derivation; in this example we obtain a maximum dynamic range of 53 dB.

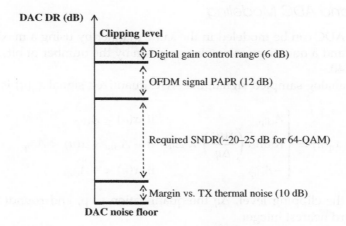

Figure 2.46 Transmitter DAC dynamic range derivation

In reception the ADC dynamic range is more complicated to specify as it depends on the analog AGC precision, the residual DC offset, channel properties such as fading due to multipath propagation, and the interferers' level at the ADC input. In Figure 2.47 we give a graphical decomposition of an ADC dynamic range; in this case we obtain around 65 dB, which is 12 dB higher than the DAC dynamic range estimated above, that is, 2 bits more, for the same SNDR requirement.

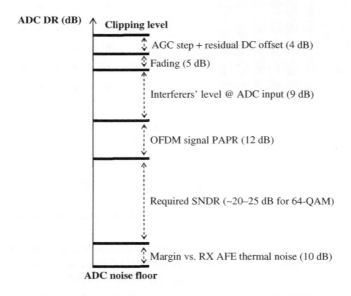

Figure 2.47   Receiver ADC dynamic range derivation

2.8.4 DAC and ADC Modeling

The DAC and ADC can be modeled in the same manner by using a maximum value, clipping level, and a quantization resolution given by the number of bits. Please refer to Equation 2.140.

If $x(n)$ is an analog sampled signal, the ADC quantized signal $x_q(n)$ is obtained by applying

$$x_q(n) = \begin{cases} A_{\text{clip}} & \text{if } x(n) \geq A_{\text{clip}} \\ \text{round}\left(\dfrac{x(n)}{\Delta q}\right) \times \Delta q & \text{if } -A_{\text{clip}} \leq x(n) < A_{\text{clip}} \\ -A_{\text{clip}} & \text{if } x(n) < -A_{\text{clip}} \end{cases} \tag{2.150}$$

where A_{clip} is the clipping level, Δq the quantization step, and round(x) a function rounding toward nearest integer.

Regarding the DAC modeling, the analog sampled signal is obtained by applying

$$x(n) = \begin{cases} A_{\text{clip}} & \text{if } x_q(n) \times \Delta q \geq A_{\text{clip}} \\ x_q(n) \times \Delta q & \text{if } -A_{\text{clip}} \leq x_q(n) \times \Delta q < A_{\text{clip}} \\ -A_{\text{clip}} & \text{if } x_q(n) \times \Delta q < -A_{\text{clip}} \end{cases} \tag{2.151}$$

2.9 IP2 and IP3: Second- and Third-Order Nonlinearities

2.9.1 Harmonics (Single-Tone Test)

If a monochromatic signal is fed into a nonlinear system, the output is composed of the input frequency (fundamental) and harmonics that are integer multiples of the fundamental frequency, as depicted in Figure 2.48. By only considering second- and third-order nonlinearities of the system, the output of a nonlinear block can be modeled as a simple polynomial function:

$$y(t) = \alpha_1 x(t) + \alpha_2 x^2(t) - \alpha_3 x^3(t) \tag{2.152}$$

where α_1 is the small-signal gain and α_2 and α_3 are positive factors related to the second- and third-order nonlinearities, respectively.

The negative term in front of α_3 is due to the fact that the third-order term compresses the gain of the fundamental and to avoid any confusion arising from the assumption $\alpha_3 < 0$.

If $x(t) = A\cos(\omega_1 t)$, we can develop Equation 2.152:

$$y(t) = \frac{\alpha_2 A^2}{2} + \left(\alpha_1 A - \frac{3\alpha_3 A^3}{4}\right)\cos(\omega_1 t) + \frac{\alpha_2 A^2}{2}\cos(2\omega_1 t) - \frac{\alpha_3 A^3}{4}\cos(3\omega_1 t)$$

$$= \text{DC} + \text{H1}\cos(\omega_1 t) + \text{H2}\cos(2\omega_1 t) + \text{H3}\cos(3\omega_1 t) \tag{2.153}$$

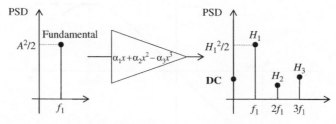

Figure 2.48 Input and output (frequency domain) of a nonlinear system (single-tone test)

in which we find a DC component, the input frequency called the fundamental with amplitude H1, and the second- and third-order terms called the "harmonics" with amplitudes H2 and H3, respectively.

As the input signal amplitude increases, the output level of the fundamental frequency varies due to the third-order term as expected.

It is easy to calculate the input amplitude for which we reach the output 1 dB compression point that is often used by RF analog designers as a metric to define the amplifier linearity:

$$20\log_{10}\left(\alpha_1 - \frac{3\alpha_3 A^2_{\text{in-1 dB}}}{4}\right) = 20\log_{10}(\alpha_1) - 1 = 20\log_{10}(10^{-1/20}\alpha_1) \qquad (2.154)$$

from which we find

$$A^2_{\text{in-1 dB}} = 0.145\left[\frac{\alpha_1}{\alpha_3}\right] \qquad (2.155)$$

or in terms of input power

$$P_{\text{in-1 dB}} = \frac{A^2_{\text{in-1 dB}}}{2} = 0.0725\left[\frac{\alpha_1}{\alpha_3}\right] \qquad (2.156)$$

Note: $P_{\text{in-1 dB}}$ is expressed here in linear, not in decibels.

Due to the third-order nonlinearity term, the output y starts to decrease as the input level increases beyond an amplitude $|x|$ that we will call $A_{\text{in-sat}}$ as depicted in Figure 2.49. Let us calculate $P_{\text{in-sat}}$ corresponding to the input amplitude $|x| = A$ for which the derivative of the fundamental defined in Equation 2.153 equals zero:

$$\frac{d}{dA}\left(\alpha_1 A - \frac{3\alpha_3 A^3}{4}\right) = \alpha_1 - \frac{9\alpha_3 A^2}{4} = 0 \qquad (2.157)$$

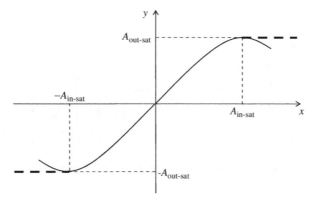

Figure 2.49   To avoid a decrease of the gain, a saturation point has to be added in the polynomial model

Consequently we find

$$A = A_{\text{in-sat}} = \pm\sqrt{\frac{4}{9}\left[\frac{\alpha_1}{\alpha_3}\right]} \tag{2.158}$$

Then for a sine wave

$$P_{\text{in-sat}} = \frac{A_{\text{in-sat}}^2}{2} = \frac{2}{9}\left[\frac{\alpha_1}{\alpha_3}\right] \tag{2.159}$$

As a result in simulation we assume for the third-order polynomial model:

$$\begin{cases} y = y(A_{\text{in-sat}}) & \text{if } x \geq A_{\text{in-sat}} \\ y = -y(A_{\text{in-sat}}) & \text{if } x \leq -A_{\text{in-sat}} \end{cases} \tag{2.160}$$

The difference between $P_{\text{in-sat}}$ and $P_{\text{in-1 dB}}$ can be expressed by their ratio in decibels:

$$10\log_{10}\left(\frac{P_{\text{in-sat}}}{P_{\text{in-1 dB}}}\right) \approx 4.9 \text{ dB} \tag{2.161}$$

Harmonics are often expressed in dBc, which is simply the ratio between the harmonic power and the fundamental power in decibels:

$$H2_{\text{dBc}} = 10\log_{10}\left(\frac{P_{H2}}{P_{H1}}\right) = 20\log_{10}\left(\left|\frac{H2}{H1}\right|\right)$$

$$H3_{\text{dBc}} = 10\log_{10}\left(\frac{P_{H3}}{P_{H1}}\right) = 20\log_{10}\left(\left|\frac{H3}{H1}\right|\right) \tag{2.162}$$

The DC power related to the second-order distortion is 3 dB above $H2_{\text{dBc}}$. Indeed, the DC level is equivalent to the amplitude of the second harmonic, which gives a factor of 2 in terms of power.

2.9.2 Intermodulation Distortion (Two-Tone Test)

Intermodulation distortion (IMD) is generally measured by applying two pure sinusoids (two-tone test), having the same power, at the circuit input. The nonlinearities of the circuit will create harmonics, described in Section 2.9.1, and combinations of the two frequencies called IMDs, as illustrated in Figure 2.50.

If we apply two equal-amplitude tones, that is, $x(t) = A\cos(\omega_1 t) + A\cos(\omega_2 t)$, to the polynomial function (2.152), we find many more components than for the single-tone test:

$$\begin{aligned} y(t) = {}& \alpha_2 A^2 + \left(\alpha_1 A - \frac{9\alpha_3 A^3}{4}\right)\cos(\omega_1 t) + \left(\alpha_1 A - \frac{9\alpha_3 A^3}{4}\right)\cos(\omega_2 t) \\ & + \frac{\alpha_2 A^2}{2}\cos(2\omega_1 t) + \frac{\alpha_2 A^2}{2}\cos(2\omega_2 t) + \alpha_2 A^2 \cos[(\omega_1 \pm \omega_2)t] \\ & - \frac{\alpha_3 A^3}{4}\cos(3\omega_1 t) - \frac{\alpha_3 A^3}{4}\cos(3\omega_2 t) \\ & - \frac{3\alpha_3 A^3}{4}\cos[(2\omega_1 \pm \omega_2)t] - \frac{3\alpha_3 A^3}{4}\cos[(2\omega_2 \pm \omega_1)t] \end{aligned} \tag{2.163}$$

As expected, in addition to DC, to the fundamental frequencies ω_1 and ω_2 and the harmonics H2 and H3, second- and third-order IMDs IMD2 and IMD3, respectively, appear:

$$\text{IMD2}(t) = \alpha_2 A^2 \cos[(\omega_1 \pm \omega_2)t]$$

$$\text{IMD3}(t) = -\frac{3\alpha_3 A^3}{4} \cos[(2\omega_1 \pm \omega_2)t] - \frac{3\alpha_3 A^3}{4} \cos[(2\omega_2 \pm \omega_1)t]$$

(2.164)

As for the harmonics, it is customary to express IMD2 and IMD3 in dBc relative to the level of the fundamental H1:

$$\text{IMD2}_{\text{dBc}} = 20 \log_{10} \left(\left| \frac{\text{IMD2}}{\text{H1}} \right| \right) = \text{H2}_{\text{dBc}} + 6$$

$$\text{IMD3}_{\text{dBc}} = 20 \log_{10} \left(\left| \frac{\text{IMD3}}{\text{H1}} \right| \right) = \text{H3}_{\text{dBc}} + 9.5$$

(2.165)

H2_{dBc} and H3_{dBc} being the second and third harmonic levels, respectively, also in dBc relative to the fundamental level H1.

More commonly the two-tone nth-order IMD IMDn of a circuit is expressed with the nth-order intercept point IPn, which is the "virtual" point where the nth-order term IMDn (as extrapolated from small-signal conditions) crosses the extrapolated power of the fundamental. This is illustrated in Figure 2.51, in which IIPn is the nth-order input intercept point and OIPn is the nth-order output intercept point. The asymptotic slope of the fundamental H1 is 1 dB/dB, whereas the asymptotic slope of the nth-order product is n dB/dB. Consequently, we can express IPn as a function of IMDn power

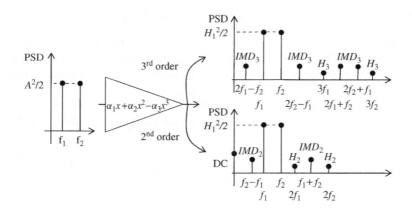

Figure 2.50   IMD produced by two pure sinusoids (two-tone test)

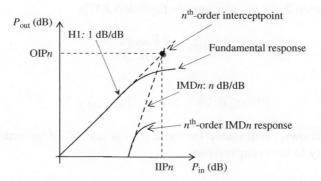

Figure 2.51 The *n*th-order intercept point IP*n* is the point where the *n*th-order term IMD*n* (as extrapolated from small-signal conditions) crosses the extrapolated power of the fundamental

$P_{\text{IMD}n}$ and the power of the fundamental P_{H1} by writing

$$\text{IP}n = P_{\text{H1}} + \Delta P = P_{\text{IMD}n} + n\Delta P \Rightarrow \Delta P = \frac{P_{\text{H1}} - P_{\text{IMD}n}}{n-1} \tag{2.166}$$

ΔP being the difference in decibels between the power P_{H1} of the fundamental H1 and the power $P_{\text{IMD}n}$ of the undesired *n*th-order product IMD*n*.

After some operations, we obtain

$$\text{IP}n = P_{\text{H1}} + \frac{P_{\text{H1}} - P_{\text{IMD}n}}{n-1} = \frac{nP_{\text{H1}} - P_{\text{IMD}n}}{n-1} \tag{2.167}$$

which gives, in decibels, for the second- and third-order intercept points:

$$\text{IP2} = 2P_{\text{H1}} - P_{\text{IMD2}}$$
$$\text{IP3} = \frac{3P_{\text{H1}} - P_{\text{IMD3}}}{2} \tag{2.168}$$

Let us express IP2 and IP3 as a function of the coefficients α_1, α_2, and α_3:

$$\alpha_1 A_{\text{IP2}} = \alpha_2 A_{\text{IP2}}^2 \Rightarrow \text{IP2} = \frac{A_{\text{IP2}}^2}{2} = \frac{1}{2}\left[\frac{\alpha_1}{\alpha_2}\right]^2 \tag{2.169}$$

and

$$\alpha_1 A_{\text{IP3}} = \frac{3\alpha_3 A_{\text{IP3}}^3}{4} \Rightarrow \text{IP3} = \frac{A_{\text{IP3}}^2}{2} = \frac{2}{3}\left[\frac{\alpha_1}{\alpha_3}\right] \tag{2.170}$$

By using Equation 2.156 we can rewrite Equation 2.170:

$$\text{IP3} = \frac{2}{3}\frac{P_{\text{in-1 dB}}}{0.0725} \tag{2.171}$$

or in decibels:

$$10\log_{10}(\text{IP3}) = 10\log_{10}(P_{\text{in-1 dB}}) + 9.6 \tag{2.172}$$

which is a well-known relationship between $P_{\text{in-1 dB}}$ and IP3 if we only consider third order nonlinearity in the compression.

2.9.3 Receiver Performance Degradation due to the Non-linearities

The receiver nonlineraties can be responsible of out-of band interferers aliasing into the signal bandwidth and corrupting the system performance. Consequently, it is the responsibility of the system designer to specify the receiver IP2 and IP3 in order to limit the impact of these impairments.

2.9.3.1 Impact of the Second-Order Nonlinearity

The second-order nonlinearity distortions are illustrated in Figure 2.52. Usually this impairment affects the signal after frequency down-conversion to the baseband, that is, after the mixer. We distinguish two cases:

- The intermodulation of two interferers generating distortion in baseband because their frequency spacing $f_2 - f_1$ is smaller than the baseband bandwidth (top of Figure 2.52).
- A single interferer having non-constant envelope, in which case the second-order nonlinearity creates an amplitude modulation around DC (bottom of Figure 2.52).

The level of the second-order distortion is commonly specified using the two-tone test IP2 formula in decibels from Equation 2.168:

$$P_{\text{IMD2}} = 2P_{\text{int}} - \text{IP2} \tag{2.173}$$

where P_{int} is the power of the interferer and P_{IMD2} is the power of the IMD2 aliased into the band of interest.

2.9.3.2 Impact of the Third-Order Nonlinearity

The third-order nonlinearity distortion is illustrated in Figure 2.53. Contrary to the second-order distortion, it affects the signal directly at RF. An intermodulation test is generally part of the receiver specification because of the requirement to coexist with adjacent and alternate channels.

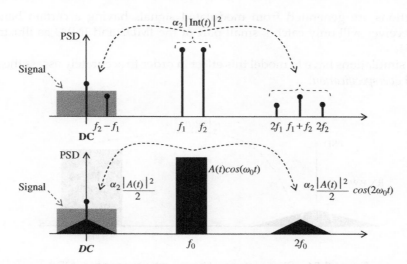

Figure 2.52 Receiver SNDR degradation due to out-of-band interferers aliased by second-order nonlinearity

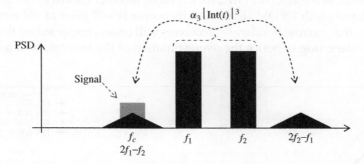

Figure 2.53 Receiver SNDR degradation due to out-of-band interferers aliased by third-order nonlinearity

The level of the third-order distortion is also commonly specified using the two-tone test IP3 formula in decibels from Equation 2.168:

$$P_{\text{IMD3}} = 3P_{\text{int}} - 2 \times \text{IP3} \qquad (2.174)$$

2.9.3.3 Two-Tone Test Specification Limitation

The classical formulas used for IP3 and IP2 specification, Equations 2.173 and 2.174, are only valid for intermodulation generated by tones as described in Section 2.9.2; they do not take into account the spectral spreading of a real signal. Indeed, in practical systems the distortions are generated from modulated signals having a certain bandwidth and the receiver will only catch a small part of the IMD2 and IMD3 as illustrated in Figure 2.54.

System simulations have to model this effect in order to accurately assess the impact, and avoid *overspecification*!

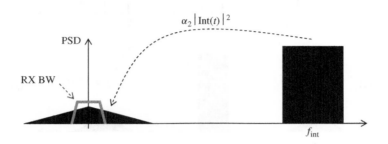

Figure 2.54   RX baseband filtering attenuates the IMDs

Figure 2.55 shows simulation results concerning SNR degradation, that is, the increase of the noise level, as a function of IIP2 for different receiver bandwidths. The interferer is a 20 MHz bandwidth OFDM signal and its power is −35 dBm at the receiver input. We clearly see that narrow bandwidth receivers will be less impacted by the distortion, and it is also interesting to notice the overestimation of the two-tone test formula. This

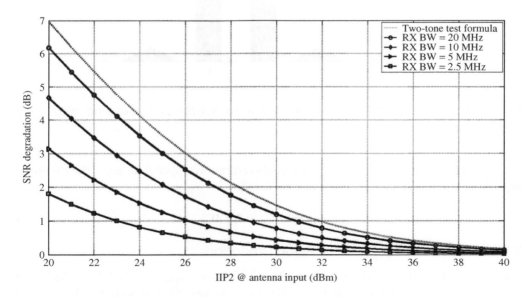

Figure 2.55   RX SNR degradation as a function of IIP2 and baseband bandwidth

kind of simulation is necessary for studying the coexistence between different systems having different bandwidths.

2.9.4 Impact of Third-Order Nonlinearity in OFDM

In the previous section we studied the receiver performance degradation due to the impact of the nonlinearities on out-of-band interferers generating distortions into the band of interest. In this section we will specifically focus on the effects of the third-order nonlinearity directly on the OFDM signal itself because it will create self-interference.

Let us consider an OFDM symbol composed of N subcarriers each having the same power P_{sub}; in this case the total power of the symbol is N times higher than the subcarrier power giving in decibels:

$$P_{in} = P_{symbol} = P_{sub} + 10\log_{10}(N) \tag{2.175}$$

Thus, the IP3 is calculated relative to the total input power:

$$IP3 = P_{in} - \frac{IMD3_{dBc}}{2} = P_{in} - \frac{H3_{dBc} + 9.5}{2} \tag{2.176}$$

For an OFDM signal, the in-band third-order distortion is specified as a composite triple beat (CTB) intermodulation derived from a three-tone test (Figure 2.56). Using the third-order polynomial function defined in Equation 2.152, it is easy to demonstrate that the CTB level is 6 dB higher than that of the IMD3 generated by two subcarriers (two-tone test):

$$CTB_{dBc} = IMD3_{dBc} + 6 \tag{2.177}$$

When more than three carriers are present in the channel, the total signal interference is related to the number of carriers and the spectral position n of the carrier as illustrated in Figure 2.57. For an OFDM signal having N equal-amplitude subcarriers, the CTB

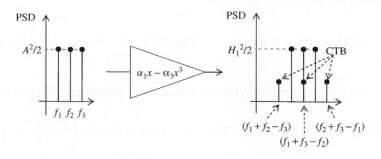

Figure 2.56 CTBs coming from third-order nonlinearity applied on three carriers

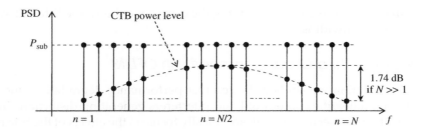

Figure 2.57   In-band CTB for an OFDM signal having N subcarriers

overlap, adding non-coherently (i.e., in power), and the total interference for subcarrier n is equal to (in dBc):

$$\text{CTB}_{\text{dBc}}(n) = 2\,(P_{\text{sub}} - IP3) + 6 + 10\log_{10}(\text{beats}(n)) \tag{2.178}$$

where

$$\text{beats}(n) = \frac{N^2}{4} + \frac{(N-n)(n-1)}{2} \tag{2.179}$$

is the number of interference carriers (beats) affecting subcarrier n ($1 \leq n \leq N$, $n = 1$ is the lowest frequency subcarrier) for N equally spaced subcarriers.

By using the total power of the symbol P_{in} instead of P_{sub} in Equation 2.178, we obtain

$$\text{CTB}_{\text{dBc}}(n) = 2(P_{\text{in}} - 10\log_{10}(N) - IP3) + 6 + 10\log_{10}\left[\frac{N^2}{4} + \frac{(N-n)(n-1)}{2}\right]$$

$$= \text{IMD3}_{\text{dBc}} - 10\log_{10}(N^2) + 6 + 10\log_{10}\left[\frac{N^2}{4} + \frac{(N-n)(n-1)}{2}\right] \tag{2.180}$$

Equation 2.180 demonstrates that in-band CTB has minimum power at the first and last subcarrier positions, that is, for $n = 1$ and $n = N$:

$$\text{CTB}_{\text{dBc}}(1) = \text{CTB}_{\text{dBc}}(N) = \text{IMD3}_{\text{dBc}} - 10\log_{10}(N^2) + 6 + 10\log_{10}\left(\frac{N^2}{4}\right) = \text{IMD3}_{\text{dBc}} \tag{2.181}$$

and the in-band *CTB* has maximum power at the center of the band, that is, for $n = N/2$:

$$\text{CTB}_{\text{dBc}}(N/2) = \text{IMD3}_{\text{dBc}} - 10Log_{10}(N^2) + 6 + 10\log_{10}\left[\frac{N(3N-2)}{8}\right]$$

$$= \text{IMD3}_{\text{dBc}} + 6 + 10\log_{10}\left(\frac{3N-2}{8N}\right) \tag{2.182}$$

If the number of subcarriers is much bigger that 1, that is, $N \gg 1$, we can estimate the difference between the minimum and the maximum CTB:

$$\text{CTB}_{\text{dBc}}(N/2) - \text{CTB}_{\text{dBc}}(1) \approx 6 + 10\log_{10}\left(\frac{3}{8}\right) = 1.74 \text{ dB} \tag{2.183}$$

Here we did not take into account the CTB components falling out-of-band, which are instead related to the out-of-band emissions.

2.9.4.1 SNDR Limitation due to the CTB

The linear CTB power level on each subcarrier n is given by

$$\text{CTB}(n) = 10^{P_{\text{sub}}/10} \times 10^{\text{CTB}_{\text{dBc}}(n)/10} = \frac{10^{P_{\text{in}}/10}}{N}\frac{10^{(\text{IMD3}_{\text{dBc}}+6)/10}}{N^2}\left[\frac{N^2}{4} + \frac{(N-n)(n-1)}{2}\right] \tag{2.184}$$

and the total CTB power within the OFDM channel is obtained by summing all the CTB components from 1 to N:

$$\text{CTB}_{\text{Total}} = \sum_{n=1}^{N}\text{CTB}(n) = \frac{10^{(P_{\text{in}}+\text{IMD3}_{\text{dBc}}+6)/10}}{N^3}\sum_{n=1}^{N}\left[\frac{N^2}{4} + \frac{(N-n)(n-1)}{2}\right] \tag{2.185}$$

By knowing the results of these two sums

$$\sum_{n=1}^{N}n = \frac{N(N+1)}{2}$$

$$\sum_{n=1}^{N}n^2 = \frac{N(N+1)(2N+1)}{6} \tag{2.186}$$

we can resolve the summation in Equation 2.185:

$$\sum_{n=1}^{N}\left[\frac{N^2}{4} + \frac{(N-n)(n-1)}{2}\right] = \frac{N^3}{3} - \frac{N^2}{4} + \frac{N}{6} \tag{2.187}$$

giving for the total power of the in-band CTB:

$$\begin{aligned}\text{CTB}_{\text{Total}} &= \frac{10^{(P_{\text{in}}+\text{IMD3}_{\text{dBc}}+6)/10}}{N^3}\left(\frac{N^3}{3} - \frac{N^2}{4} + \frac{N}{6}\right) \\ &= 10^{(P_{\text{in}}+\text{IMD3}_{\text{dBc}}+6)/10}\left(\frac{1}{3} - \frac{1}{4N} + \frac{1}{6N^2}\right)\end{aligned} \tag{2.188}$$

If the number of subcarriers is large, that is, $N \gg 1$, it is interesting to notice that CTB in-band power no longer depends on the number of subcarriers N:

$$CTB_{Total} \approx \frac{10^{(P_{in}+IMD3_{dBc}+6)/10}}{3} \tag{2.189}$$

By using Equation 2.176, we obtain

$$CTB_{Total} \approx \frac{10^{[3P_{in}-(2 \times IP3)+6]/10}}{3} \tag{2.190}$$

The SNDR is then simply the ratio of the OFDM signal power to the in-band CTB total power:

$$SNDR \approx \frac{10^{P_{in}/10}}{CTB_{Total}} = \frac{3}{10^{[2(P_{in}-IP3)+6]/10}} \approx \frac{1}{10^{[2(P_{in}-IP3)+1.2]/10}} \tag{2.191}$$

The result of Equation 2.191 is very useful for rapidly estimating the SNDR of the system as a function of the OFDM signal power and the IP3.

2.9.5 Simulation in Complex Baseband

Generally, simulations are done in baseband using complex signals instead of real signals. Consequently, the two-tone test is done using as input

$$x(t) = A\, e^{j\omega_1 t} + A\, e^{j\omega_2 t} \tag{2.192}$$

and the polynomial function modeling the second- and third-order distortions becomes

$$y(t) = \alpha_1 x(t) + \alpha_2 \left|x(t)\right|^2 - \alpha_3 x(t) \left|x(t)\right|^2 \tag{2.193}$$

If we expand the polynomial in Equation 2.193 using

$$\begin{aligned}
\left|x(t)\right|^2 &= x(t)x(t)^* = A^2 \left[2 + e^{j(\omega_1-\omega_2)t} + e^{j(\omega_2-\omega_1)t}\right] \\
x(t)\left|x(t)\right|^2 &= A^3 \left[3e^{j\omega_1 t} + 3e^{j\omega_2 t} + e^{j(2\omega_1-\omega_2)t} + e^{j(2\omega_2-\omega_1)t}\right]
\end{aligned} \tag{2.194}$$

we can easily derive the IP2 and IP3:

$$\alpha_1 A_{IIP2} = \alpha_2 A_{IIP2}^2 \Rightarrow IIP2 = A_{IIP2}^2 = \left[\frac{\alpha_1}{\alpha_2}\right]^2 \tag{2.195}$$

$$\alpha_1 A_{IIP3} = \alpha_3 A_{IIP3}^3 \Rightarrow IIP3 = A_{IIP3}^2 = \left[\frac{\alpha_1}{\alpha_3}\right] \tag{2.196}$$

Note: If we want to reuse the same IP2 and IP3 values derived from real signals (Equations 2.169 and 2.170), the complex baseband polynomial has to be rewritten as

$$y(t) = \alpha_1 x(t) + \sqrt{2}\alpha_2 |x(t)|^2 - \frac{3}{2}\alpha_3 x(t) |x(t)|^2 \qquad (2.197)$$

2.10 Power Amplifier Distortion

PAs present impairments which have to be taken into account during the system design because they can degrade the modulation accuracy (EVM) and also produce unwanted out-of-band emissions, due to spectral regrowth, causing interference to neighboring channels (Liang *et al.*, 1999; Zhou and Kenney, 2002).

The distortions of the PA are generally modeled using AM/AM (amplitude to amplitude) and AM/PM (amplitude to phase) curves which can be measured or represented as mathematical functions. If we consider a complex baseband signal $x(t) = a(t)e^{j\varphi(t)}$, the output of the PA can be written

$$y_{PA}(t) = AM(a(t))\, e^{j[\varphi(t)+PM(a(t))]} \qquad (2.198)$$

where $AM(a(t))$ is the AM/AM function describing the PA output amplitude as a function of the input signal amplitude, and $PM(a(t))$ is the AM/PM function describing the PA output phase as a function of the input signal amplitude. The AM/AM causes amplitude distortion whereas AM/PM introduces phase shift.

2.10.1 PA Modeling

The aim of the PA modeling is to develop mathematical functions for including AM/AM and AM/PM characteristics in the system simulations. In this section we will briefly describe the main models used in the literature.

Note that the models presented below do not depend on frequency and are called memory-less because their outputs do not depend on the past, only on the instantaneous input. However, with wideband communication systems memory effects can be observed in the PA and new modeling techniques based on Volterra series or memory polynomial models can be used.

2.10.1.1 Saleh Model

This model (Saleh, 1981) was originally developed for traveling-wave tube (TWT) amplifiers:

$$AM(a(t)) = \frac{\alpha_G |x(t)|}{1 + \beta_G |x(t)|^2}$$

$$PM(a(t)) = \frac{\alpha_\phi |x(t)|^2}{1 + \beta_\phi |x(t)|^2} \qquad (2.199)$$

The original coefficients proposed for TWT amplifiers modeling are

$$\alpha_G = 2.1587,\ \beta_G = 1.1517,\ \alpha_\phi = 4.033,\ \beta_\phi = 9.1040.$$

For solid-state power amplifier (SSPA) modeling, Rapp (1991) proposed another set of coefficients:

$$\alpha_G = 1,\ \beta_G = 0.25,\ \alpha_\phi = 0.26,\ \beta_\phi = 0.25.$$

2.10.1.2 Ghorbani Model

This model (Ghorbani and Sheikhan, 1991) is quite similar to Saleh's, but more focused on SSPA modeling.

$$\text{AM}(a(t)) = \frac{x_1 \left|x(t)\right|^{x_2}}{1 + x_3 \left|x(t)\right|^{x_2}} + x_4 \left|x(t)\right|$$

$$\text{PM}(a(t)) = \frac{y_1 \left|x(t)\right|^{y_2}}{1 + y_3 \left|x(t)\right|^{y_2}} + y_4 \left|x(t)\right| \tag{2.200}$$

For a GaAs field-effect transistor PA, Ghorbani proposed this set of coefficients:

$$x_1 = 8.1081,\ x_2 = 1.5413,\ x_3 = 6.5202,\ x_4 = -0.0718$$
$$y_1 = 4.6645,\ y_2 = 2.0965,\ y_3 = 10.88,\ y_4 = -0.003$$

2.10.1.3 Rapp Model

Contrary to the two previous models, the Rapp model (Rapp, 1991) deals only with AM/AM characteristics in order to produce a smooth saturation of the envelope transfer characteristic:

$$\text{AM}(a(t)) = \frac{G \left|x(t)\right|}{\left[1 + \left(\dfrac{G \left|x(t)\right|}{V_{\text{sat}}}\right)^{2p}\right]^{1/2p}}$$

$$\text{PM}(a(t)) = 0 \tag{2.201}$$

in which p is the smoothness factor dictating how the PA reaches saturation, V_{sat} is the saturation level, and G is the small-signal gain.

Rapp proposed using $p = 3$ for accurate modeling of SSPA.

In order to rapidly visualize these three PA models, we plotted in Figures 2.58 and 2.59 their AM/AM and AM/PM characteristics, respectively.

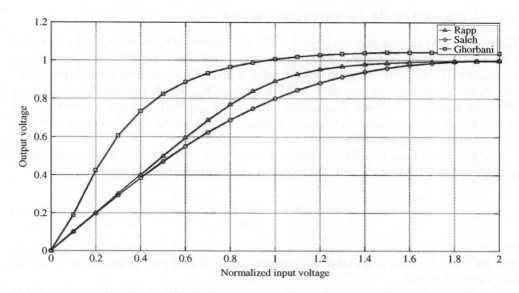

Figure 2.58 AM/AM characteristics for Rapp, Saleh, and Ghorbani models

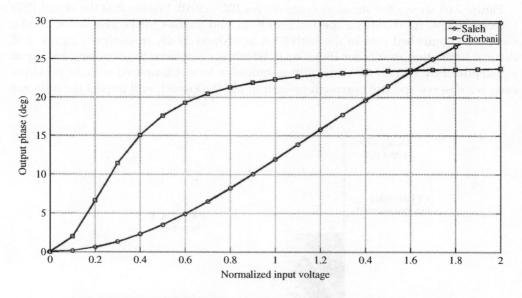

Figure 2.59 AM/PM characteristics for Saleh and Ghorbani models

2.10.2 Impact of PA Distortions in OFDM

Because OFDM signals present large PAPRs as detailed in Section 2.8.2, OFDM signal PAs need to be very linear in order to limit signal distortions and out-of-band emissions (Costa *et al.*, 1999; Dardari *et al.*, 2000). PA saturation is like a softer clipping "squeezing" the signal amplitude as opposed to DAC and ADC, which present hard clipping. Whereas constant-envelope modulations like GSM use high-efficiency switching PA (Class E), OFDM modulation requires a linear PA (Class A) with large back-off and consequently very poor efficiency.

The PA input back-off (IBO) is defined as the ratio between the 1 dB compression point input-referred power and the input signal average power:

$$\text{IBO} = \frac{P_{\text{in-1 dB}}}{\left\langle \left| x(t) \right|^2 \right\rangle} \qquad (2.202)$$

The IBO specification is illustrated in Figure 2.60 by highlighting the influence of the large OFDM PAPR.

The two following figures show the impact of the PA back-off on a 64-QAM constellation and SNR for two different cases using the Rapp model.

Figure 2.61 shows the simulation results for IBO = 9 dB. We see that the signal PSD is clean, that is, it exhibits no spectral regrowth, and the 64-QAM constellation can be clearly distinguished due to the high SNR of around 37 dB. In contrast, Figure 2.62 depicts simulation results for IBO = 6 dB, which is equivalent to increasing the input signal power by 3 dB for the same output saturation level. Compared with the previous case, we observe spectral regrowth present in the PSD which will impact the adjacent

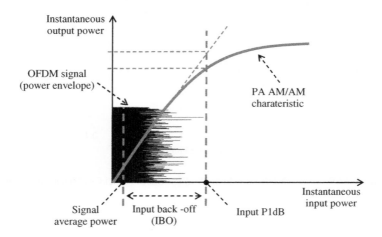

Figure 2.60   PA input back-off with OFDM signal

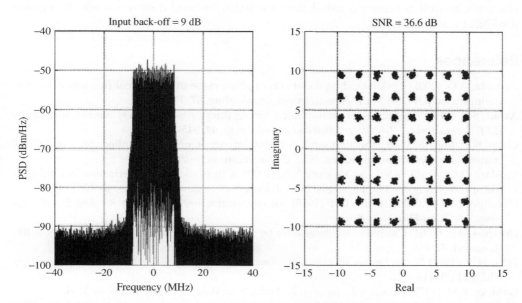

Figure 2.61 OFDM signal PSD and 64-QAM constellation at PA output with 9 dB input back-off

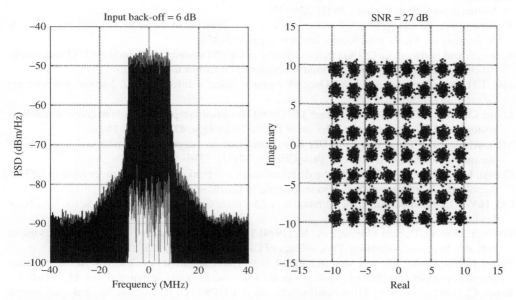

Figure 2.62 OFDM signal PSD and 64-QAM constellation at PA output with 6 dB input back-off

channels, as well as noisier constellation due to the in-band distortions which degrade the SNR by around 10 dB.

References

Armada, A.G. (2001) Understanding the effects of phase noise in orthogonal frequency division multiplexing (OFDM). *IEEE Transactions on Broadcasting*, **47**, 153–159.

Awad, S.S. (1995) The effects of accumulated timing jitter on some sine wave measurements. *IEEE Transactions on Instrumentation and Measurement*, **44**, 945–951.

Costa, E.E., Midrio, M.M., and Pupolin, S.S. (1999) Impact of amplifier nonlinearities on OFDM transmission system performance. *IEEE Communications Letters*, **3**, 37–39.

Dardari, D.D., Tralli, V.V., and Vaccari, A.A. (2000) A theoretical characterization of nonlinear distortion effects in OFDM systems. *IEEE Transactions on Communications*, **48**, 1755–1763.

Davenport, W.B. and William, L.R. (1958) *An Introduction to the Theory of Random Signals and Noise*, McGraw-Hill, New York.

Drakhlis, B.B. (2001) Calculate oscillator jitter by using phase-noise analysis. *Microwaves & RF*, (January), 82–90.

Friis, H.T. (1944) Noise figures of radio receivers. *Proceedings of the Institute of Radio Engineers (IRE)*, **32** (7), 419–422.

Gardner, F.M. (1979) *Phaselock Techniques*, 2nd edn, John Wiley & Sons, Inc., New York.

Ghorbani A. and Sheikhan M. (1991) The effect of solid state power amplifiers (SSPAs) nonlinearities on MPSK and M-QAM signal transmission. Proceedings of International Conference on Digital Processing of Signals in Communications, pp. 193–197.

Hajimiri, A. and Lee, T.H. (1998) A general theory of phase noise in electrical oscillators. *IEEE Journal of Solid-State Circuits*, **35** (3), 326–336.

Kleinpenning, T.G.M. (1990) $1/f$ noise in electronic devices, in *Noise in Physical Systems* (ed. A. Ambrozy), AkadCmiai Kiadb, Budapest, pp. 443–454.

Krone, S. and Fettweis, G. (2008) On the capacity of OFDM systems with receiver I/Q imbalance. Proceedings of the IEEE International Conference on Communications.

Lee, T.H. and Hajimiri, A. (2000) Oscillator phase noise: a tutorial. *IEEE Journal of Solid-State Circuits*, **35** (3), 326–336.

Liang, C., Jong, J., Stark, W.E., and East, J.R. (1999) Nonlinear amplifier effects in communications systems. *IEEE Transactions on Microwave Theory Technology*, **47** (8), 1461–1466.

Mansuri, M. and Yang, C.K.K. (2002) Jitter optimization based on phase-locked loop design parameters. *IEEE Journal of Solid-State Circuits*, **37** (11), 1375–1382.

Ochiai, H. and Imai, H. (2001) On the distribution of the peak-to-average power ratio in OFDM signals. *IEEE Transactions on Communications*, **49** (2) 282–289.

Ott, H.W. (1976) *Noise Reduction Techniques in Electronic Circuits*, John Wiley & Sons, Inc., New York.

Pollet, T., Spruyt, P., and Moeneclaey, M. (1994) The BER performance of OFDM systems using non-synchronized sampling. Proceedings of Globecom, pp. 253–257.

Pollet, T., Van Bladel, M., and Moeneclaey, M. (1995) BER sensitivity of OFDM systems to carrier frequency offset and Wiener phase noise. *IEEE Transactions on Communications*, **43**, 191–193.

Rapp, C. (1991) Effects of HPA-nonlinearity on a 4-DPSK/OFDM-signal for a digital sound broadcasting system. Proceedings of the Second European Conference on Satellite Communications.

Saleh, A.A.M. (1981) Frequency-independent and frequency-dependent nonlinear models of TWT amplifiers. *IEEE Transactions on Communications*, **29** (11), 1715–1720.

Silicon Laboratories (2006) Estimating period jitter from phase noise, Application Note AN279 (Rev. 0.1).

Vandamme, L.K.J. (1990) $1/f$ noise in CMOS transistors, in *Noise in Physical Systems*, (ed. A. Ambrozy), AkadCmiai Kiado, Budapest, pp. 491–494.

Zhou, G.T. and Kenney, J.S. (2002) Predicting spectral regrowth of nonlinear power amplifiers. *IEEE Transactions on Communications*, 718–722.

3

Simulation of the RF Analog Impairments Impact on Real OFDM-Based Transceiver Performance

3.1 Introduction

The major RF analog impairments we find in modern transceivers have been described and mathematically modeled in Chapter 2. The aim of this chapter is to address the system simulation of these imperfections in OFDM-based transceivers so as to study their impact on the RF analog transmitter and receiver performance. To summarize, here are the RF analog impairments we will take into account:

- In transmission:
  - DAC resolution, clipping, and clock sampling jitter;
  - LO phase noise;
  - quadrature imbalance;
  - PA distortions.
- In reception:
  - LNA nonlinearities IP2 and IP3;
  - quadrature imbalance;
  - LO phase noise;
  - CFO and SFO;
  - ADC resolution, clipping, and clock sampling jitter.

RF Analog Impairments Modeling for Communication Systems Simulation:
Application to OFDM-based Transceivers, First Edition. Lydi Smaini.
© 2012 John Wiley & Sons, Ltd. Published 2012 by John Wiley & Sons, Ltd.

The resulting analysis will be based on EVM estimation as a function of each of the RF analog impairments. The principle is to vary all the impairments individually and to observe their impact on the system performance through EVM estimation. In addition we will illustrate the results using PSDs, showing the noise and the distortion generated by the imperfections inside and outside the channel bandwidth. In order to target OFDM-based transceiver performance, we will compare the simulation results for two deployed standards: WLAN (OFDM WiFi: IEEE 802.11a/g) and mobile WiMAX (IEEE 802.16e). Actually, the choice of these two technologies is significant from an RF analog point of view because they use different subcarrier spacings; large for WLAN (312.5 kHz) and narrow for WiMAX (~11 kHz). Indeed, we will see later that the behavior of some imperfections (phase noise, SFO, and CFO) will depend especially on the subcarrier spacing. In addition, it is interesting to note that the results obtained for mobile WiMAX can be extrapolated to other recent standards such as 3GPP-LTE developed for 4G mobile phones.

Regarding the RF analog front-end modeling, it will be based on the zero-IF architecture, which is the most commonly used for on-chip integration. However, the results can be easily extrapolated to other architectures using non-zero IF.

3.2 WLAN and Mobile WiMAX PHY Overview

3.2.1 WLAN: Standard IEEE 802.11a/g

"WiFi" is a brand name used by the WiFi Alliance (industry consortium) certifying the interoperability between the WLAN systems specified by the IEEE 802.11 standard (IEEE, 1999, 2003). WLAN systems target high data-rate wireless communications for local networking (coverage radius of several tens of meters). The 802.11a and the 802.11g flavors are similar because they use the same OFDM modulation; the difference is the band (carrier frequency), unlicensed 5 GHz U-NII for 11a and 2.4 GHz ISM (industrial, scientific, and medical) for 11g. The channel bandwidth is 20 MHz and 48 subcarriers are used to transmit data. Four additional subcarriers are dedicated to pilots, the DC subcarrier is unused, and 11 subcarriers are blanked to provide guard bands (six on the left and five on the right), giving a total number of 64 subcarriers for the IFFT/FFT size. Pilot subcarriers are only modulated using binary phase shift keying (BPSK) whereas data can be transmitted using quadrature phase shift keying (QPSK), 16-QAM or 64-QAM depending on the channel conditions. A convolutional channel coding is specified with three coding rates: 1/2, 2/3, and 3/4.

The IFFT/FFT sampling frequency being 20 MHz and the number of subcarriers 64, the subcarrier spacing is then 20 MHz/64 = 312.5 kHz, giving a useful OFDM symbol duration of 3.2 μs. By adding a cyclic prefix of one fourth of the symbol length, that is, 0.8 μs, we obtain a total OFDM symbol duration of 4 μs.

The theoretical maximum PHY data-rate is 54 Mbits/s, obtained using 64-QAM (6 bits each on the 48 data subcarriers) and minimum coding rate of 3/4: ((6 × 48)/4 μs) × 3/4.

Table 3.1 summarizes the main WLAN PHY parameters based on the standards IEEE 802.11a and g.

Table 3.1 IEEE 802.11a/g PHY parameters

Channel bandwidth (MHz)	20
IFFT/FFT size	64
IFFT/FFT clock (MHz)	20
Subcarrier spacing (kHz)	312.5
Used subcarriers	52 (48 for data and 4 for pilots)
Guard band subcarriers	Left: 6; right: 5
Modulation	BPSK, QPSK, 16-QAM, 64-QAM
Useful OFDM symbol duration T_u (μs)	3.2
Cyclic prefix length (μs)	$1/4 \times T_u = 0.8$
Total OFDM symbol duration (μs)	4
Channel coding	Convolutional coding rates:1/2, 2/3, 3/4

3.2.2 Mobile WiMAX: Standard IEEE 802.16e

Mobile WiMAX IEEE 802.16e is the mobility mode of the WMAN system defined by the IEEE 802.16 standard (IEEE, 2005; Chen *et al.*, 2008). It is a mobile broadband wireless communication system, and part of the fourth generation of cellular wireless communication technologies commonly called 4G by the network operators. Although WiMAX was inspired by WLAN standard 802.11 and uses OFDM modulation, its PHY is quite different because it targets mobile devices moving at high speeds (several tens of kilometers per hour).

Mobile WiMAX can be deployed in licensed or unlicensed bands between 2 and 11 GHz, its PHY is based on S-OFDMA supporting different channel bandwidths from 1.25 to 20 MHz. OFDMA is a technology based on the sharing of the OFDM symbol subcarriers amongst several users. The result is a sub-channelization of the OFDM symbol and the standard defines different modes for the allocation and the permutation of the subchannels: Partial Use of the Subchannel (PUSC) or Full Use of the Subchannel (FUSC). The PHY is scalable because the number of subcarriers is based on the channel bandwidth, from 128 for 1.25 MHz to 2048 for 20 MHz, in order to keep the subcarrier spacing constant. Mobile WiMAX also supports an adaptive modulation and coding (AMC) technique allowing the optimization of the PHY parameters, namely modulation order and coding rate, to the radio link quality in order to maintain a required BER. If the SNR is good, subcarriers can be mapped with 64-QAM, whereas if the SNR is low a robust BPSK can be employed. Regarding the channel coding, 802.16e proposes different FEC: mandatory convolutional coding, required convolutional turbo codes, and optional block turbo codes and low density parity check (LDPC) codes.

Table 3.2 shows the principal OFDMA PHY parameters of mobile WiMAX in which we can see the IFFT/FFT length and the number of sub-channels varying as a function of the channel bandwidth.

Table 3.2   IEEE 802.16e (mobile WiMAX) OFDMA PHY scalability parameters

Channel bandwidth (MHz)	1.25	5	10	20
IFFT/FFT size	128	512	1024	2048
IFFT/FFT clock (bandwidth × 28/25) (MHz)	1.4	5.6	11.2	22.4
Number of subchannels	2	8	16	32
Subcarrier spacing (kHz)	10.94			
Used subcarriers	72	360	720	1440
Max. pilot subcarriers	12	60	120	240
Guard band subcarriers + DC	44	92	184	368
Modulation	BPSK, QPSK, 16-QAM, and 64-QAM			
Useful OFDM symbol duration T_u (µs)	91.4			
Cyclic prefix length	1/32, 1/16, 1/8, 1/4 × T_u			
Total OFDM symbol duration (µs)	94.29–114.29			
Channel coding	Convolutional optional convolutional turbo, block turbo, and LDPC			

3.3 Simulation Bench Overview

3.3.1 WiFi and WiMAX OFDM Transceiver Modeling

Figure 3.1 shows the block diagram of the OFDM transceiver simulator used to study the impact of the analog front-end imperfections on the system performance. All the simulations are performed using the baseband complex representation introduced and described in Section 1.4.4.

In transmission, random data symbols are generated and mapped to an *XX-QAM* complex constellation at the OFDM modulator input. The pilot-tones are modulated with fixed amplitude and phase. Afterwards, the pilot-tones and the data symbol constellations are mapped onto subcarriers in the frequency domain before applying an IFFT which generates the time-domain OFDM symbol. In the time domain a cyclic prefix is added to produce a guard interval between OFDM symbols, and finally an up-sampling by four delivers the digital signal to the transmit analog front-end DAC.

The receive path starts with the analog front-end which delivers the OFDM signal, digitized by an ADC, to the digital baseband. The down-sampling by four after the ADC is necessary to recover the IFFT/FFT sampling resolution imposed by the standard; after that the cyclic prefix is removed and an FFT is applied. Afterwards, channel estimation is performed from the pilot-tones in order to equalize the FFT output, especially to compensate the phase rotation as will be seen, before the data demodulation gives

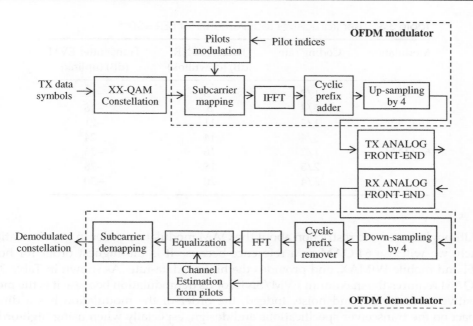

Figure 3.1   OFDM transceiver simulator

the complex constellation. For WiFi there is no implementation choice of the OFDM modulator/demodulator simulator, since the IEEE 802.11a/g PHY parameters only define a 20 MHz channel, 64-point IFFT/FFT clocked at 20 MHz, and a cyclic prefix ratio of 1/4 (Table 3.1). On the other hand, as the mobile WiMAX PHY is based on S-OFDMA technology, different channel bandwidths, and IFFT/FFT sizes are specified (Table 3.2). For our simulations we will only consider the 20 MHz channel bandwidth in order to align with the WiFi case. The main difference is the IFFT/FFT size giving a large subcarrier spacing for WiFi, 312.5 kHz, and a narrow one for mobile WiMAX, 10.94 kHz. The cyclic prefix is fixed to 1/4 for both WiFi and WiMAX giving OFDM symbol durations of 4 μs and 114.29 μs, respectively. The sampling rate used for the analog front-end simulation is imposed by choice of the DAC and ADC clocks running at four times the IFFT/FFT sampling rate, that is, 80 MHz for WiFi and 89.6 MHz for mobile WiMAX.

For the performance evaluation we will use as a reference the SNR and the EVM specified in Table 3.3 for mobile WiMAX downlink reception and uplink transmission. These specifications are the minimum requirements for guaranteeing BER = 10^{-6} in an AWGN channel using convolutional FEC. We can note that the transmission EVM requirements are 10 dB stricter than the receive SNR requirements. This results from the link budget specification and the fact that the in-band transmitter noise floor should not limit the reception performance at the sensitivity level. If, for example, the specifications were the same in transmission and reception, the overall system performance will be degraded by 3 dB.

Table 3.3   User transceiver requirements for BER $= 10^{-6}$

Modulation	Coding rate	Receiver SNR (dB) (downlink)	Transmitter EVM (dB) (uplink)
QPSK	1/2	5	−15
	3/4	8	−18
16-QAM	1/2	11	−21
	3/4	14	−24
64-QAM	1/2	16	−26
	2/3	18	−28
	3/4	20	−30

Although there is no restriction about the QAM modulation order in our simulation bench, we will use 64-QAM as a reference because it is the highest order for both WiFi and mobile WiMAX, and provides the highest data-rate. As shown in Table 3.3, 64-QAM requires the maximum EVM/SNR for its demodulation because it is the most sensitive to distortion and noise. Indeed, the order of the modulation has a direct impact on the transceiver specifications and design, especially when using high-order modulations to achieve high data-rate.

3.3.2 EVM Estimation as Performance Metric

In the lab, analog front-end performance is generally measured using a vector signal analyzer (VSA) in order to observe in real time the digitally modulated constellation and to compute the EVM as a performance metric. From a communication system point of view, EVM is the RMS value of the constellation error, that is, the difference, between the ideal transmitted symbols and the demodulated ones:

$$\text{EVM} = \sqrt{\frac{\frac{1}{N} \sum_{n=1}^{N} \left| s_{\text{RX},n} - s_{\text{TX},n} \right|^2}{\frac{1}{N} \sum_{n=1}^{N} \left| s_{\text{TX},n} \right|^2}} \tag{3.1}$$

where $s_{\text{RX},n}$ and $s_{\text{TX},n}$ are the nth RX measured symbol and the TX ideal one, respectively. Actually, EVM is commonly used as a specification and performance metric for transmitters, whereas SNR is more often used for the receiver because it comes directly from digital baseband BER vs. E_b/N_0 performance curves (Table 3.3). However, for RF analog system simulation and measurement, SNR is difficult to compute in the time domain because it is not obvious how to separate the signal from the noise, especially for OFDM modulation which looks like Gaussian noise. On the other hand, EVM is much easier to obtain because it requires only an FFT so as to extract the symbols and to compare with the ideal ones. Because we demonstrated in Section 1.4.2 that EVM and SNR are equivalent, we will also use EVM for quantifying the receiver performance.

In addition, we added in our simulation bench a spectral analysis which is very useful to observe the in-band distortions causing EVM degradation and also to measure the out-of-band emissions; for example, the ACLR.

The EVM estimation and spectral analysis simulation bench are presented in Figure 3.2, in which we can see that the ideal transmission constellation is also used in reception after the OFDM demodulation in order to compute the EVM. The spectral analysis of the distortion is done by using the OFDM signal at the input and the output of the analog front-end.

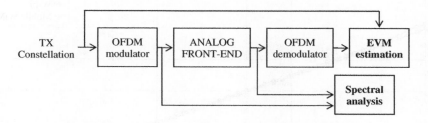

Figure 3.2 EVM estimation and spectral analysis

3.3.3 EVM versus SNR Simulations in AWGN Channel

In order to validate the EVM estimation and the spectral analysis processing of our simulation bench, we first simulated an AWGN channel configuration for which the results can be exactly predicted. Indeed, as we have seen in Section 1.4.2, the EVM is theoretically the inverse of the square root of the SNR in linear, or the negative value in decibels:

$$\mathrm{EVM_{dB}} = 20 \log_{10} \left(\frac{1}{\sqrt{\mathrm{SNR}}} \right) = -10 \log_{10}(\mathrm{SNR}) = -\mathrm{SNR_{dB}} \qquad (3.2)$$

Consequently, the aim of this preliminary simulation is to measure the EVM as a function of the OFDM signal SNR, by setting a noise level at the demodulator input, and to see if we retrieve the linear relationship defined by Equation 3.2. The AWGN simulation set-up is depicted in Figure 3.3.

Figure 3.4 shows the simulation result for both WiFi and mobile WiMAX with the SNR of the signals varying from 0 to 50 dB. The signal power has been set to 0 dBm (physical power unit in order to compute PSD in dBm/Hz as we measure in the lab) and the desired SNR is used to generate the proper AWGN level. As expected,

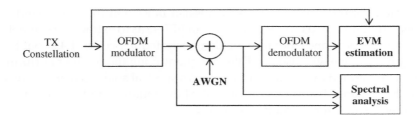

Figure 3.3   EVM estimation in AWGN channel for a given SNR at demodulator input

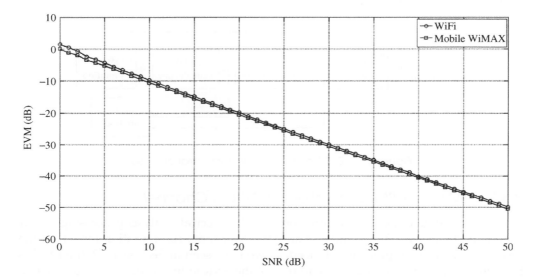

Figure 3.4   WiFi and Mobile WiMAX measured EVM as a function of input SNR

the measured EVM is the negative of the input SNR in decibels for both WiFi and WiMAX simulation benches, confirming Equation 3.2 and thus validating our EVM estimation.

Figures 3.5 and 3.6 show the demodulated complex constellations and the PSD for two specific cases: 40 dB SNR for mobile WiMAX and 35 dB SNR for WiFi, respectively. We can read from the above complex constellations that the EVM estimations are in line with the imposed input SNR. It is also interesting to note that the average PSD levels for the signal and the noise are also consistent, since 0 dBm signal power gives around −73 dBm/Hz for a 20 MHz channel bandwidth, and the noise PSD scales correctly with the SNR.

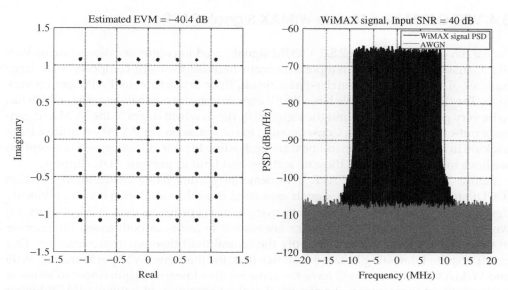

Figure 3.5 EVM estimation and PSD for WiMAX signal simulated with 40 dB SNR imposed by AWGN

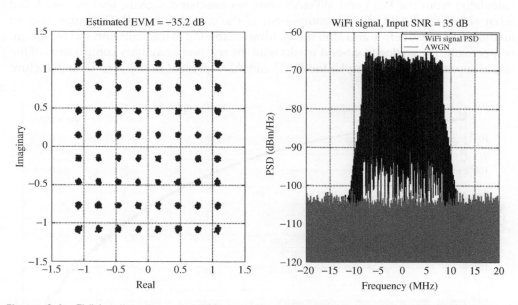

Figure 3.6 EVM estimation and PSD for WiFi signal simulated with 35 dB SNR imposed by AWGN

3.4 WiFi OFDM and Mobile WiMAX Signals PAPR

As we have seen in Section 2.8.2, OFDM signals used in communication systems look like Gaussian noise in the temporal domain because they are composed of a large number of modulated subcarriers; as a result, their instantaneous PAPR can be very high. The direct impact of this large PAPR concerns the transceiver linearity, which has to be very good in order to limit the signal distortions which degrade the EVM and generate out-of-band emissions. Consequently, in transmission the PA may require a large back-off in order to operate in its linear range, leading to poor efficiency, and in reception we have to carefully control the gain settings and limit or prevent ADC clipping.

Figure 3.7 presents the PAPR complementary cumulative distribution function (CCDF) for WiFi and WiMAX signals estimated from the simulation bench previously described. Although WiMAX employs many more subcarriers than WiFi, 2048 vs. 64, we can see that their PAPR statistics are similar because, in both cases, the number of subcarriers is large enough to apply the central limit theorem approximation. This result is quite interesting because it shows that, for the same EVM performance, WiFi and WiMAX transceivers will have the same relative linearity requirement in terms of back-off. From these curves we can see that the probability of having a PAPR higher than 8 dB is around 10^{-3}, and is lower than 10^{-7} for a PAPR superior to 12 dB. It is important to keep in mind that the PAPR CCDF curves plotted in Figure 3.7 have been calculated from the WiFi and WiMAX complex baseband signals, that is, $I + jQ$. But what about the PAPR of each component of the complex signal? This must also be analyzed, because before and after the quadrature mixing in transmission and reception, respectively, the analog front-end works with the real and imaginary components of the signal separately. For example, the DAC and ADC present in the zero-IF architecture

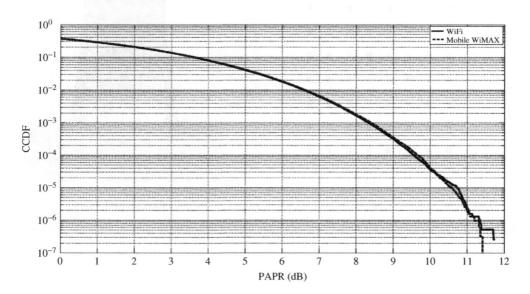

Figure 3.7   WiFi and mobile WiMAX signal PAPR CCDF

will convert each component independently as orthogonal components of the complex baseband signal.

Figure 3.8 shows the PAPR CCDF of the WiMAX complex baseband signal and those of each of its real and imaginary components. We also added the PAPR CCDF computed for a Gaussian noise for comparison. As expected, the PAPR of the signal components are similar to each other; on the other hand, the PAPR is higher than that of the complex signal. This can be explained by the fact that the complex signal is the incoherent sum of the real and imaginary parts of the OFDM signal, which are orthogonal, and consequently they add in power but not in amplitude. It is very interesting to see that the PAPR of the OFDM signal components is identical to that of Gaussian noise, confirming the assumption that an OFDM signal has a Gaussian distribution in the time domain.

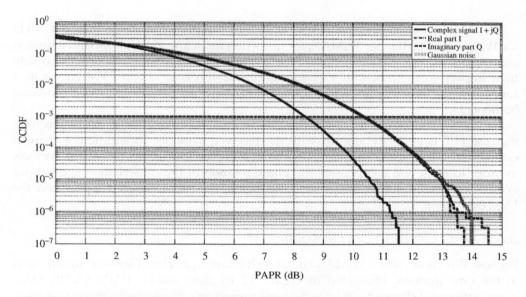

Figure 3.8 WiMAX PAPR for complex signal vs. real and imaginary components

3.5 Transmitter Impairments Simulation

3.5.1 Introduction

Figure 3.9 shows the configuration of the transmitter analog front-end simulator, including the following impairments modeling:

- DAC quantization and clipping;
- quadrature mixer IQ mismatch (phase and amplitude);
- LO phase noise ;
- PA distortion.

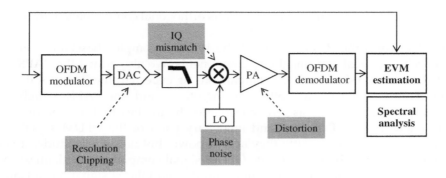

Figure 3.9   Transmitter analog front-end impairments simulated bench

The DAC is clocked at four times the IFFT/FFT sampling frequency, that is, 80 MHz for WiFi and 89.6 MHz for WiMAX, for distortion modeling and out-of-band emission measurements. We will study each of these impairments separately in order to isolate their impact on the transmitter performance; the aim is to obtain EVM curves as a function of the impairments, which allows us to easily identify the system performance bottlenecks.

For the OFDM signal demodulation, the time synchronization is ideal, meaning that the FFT is perfectly aligned with the received symbols. In addition, a channel estimation is done using the pilots present in both WiFi and WiMAX OFDM symbols in order to track and compensate the phase rotation of the constellation symbol per symbol, especially due to the phase noise, before the subcarrier demapping and the EVM estimation.

3.5.2 DAC Clipping and Resolution

Because WiFi and mobile WiMAX signals exhibit an equivalent Gaussian distribution in the time domain, illustrated in Figure 3.7 by their similar PAPR CCDF, DAC quantization and clipping will have the same impact on each. Actually, this assumption can be extrapolated to all OFDM signals having a large number of subcarriers, in so much as they resemble Gaussian noise in the temporal domain. Figure 3.10 shows the transmitter EVM as a function of the DAC clipping ratio and number of bits. The clipping ratio is merely defined as the DAC saturation level divided by the signal RMS value or equivalently their difference in decibels; here, it varies from 6 to 18 dB (1–3 bits). Regarding the DAC resolution, we considered 8, 9, 10, and 11 bits. For a clipping ratio lower than 10 dB the clipping noise dominates and we see that there is no advantage in improving the DAC resolution, as the EVM consistently exceeds −40 dB. However, above an 11 dB clipping ratio we start to observe the impact of the quantization noise and the effect of the DAC resolution. For an 8-bit DAC, we clearly notice that increasing the clipping ratio above 12 dB does not help; and even worse, the EVM is degraded because the signal level is diminishing while the quantization

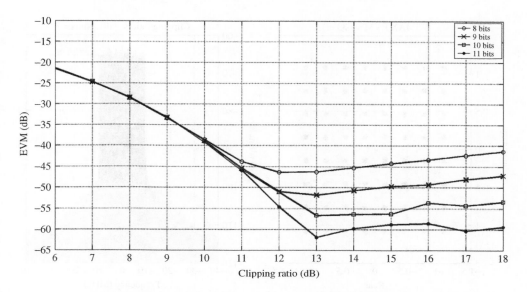

Figure 3.10 WiMAX transmitter EVM as a function of DAC clipping ratio and resolution

noise level remains fixed. The impact of the DAC resolution is particularly noticeable for clipping ratios above 13 dB, when the clipping noise starts to become negligible compared with the quantization noise, and in this case we recover 6 dB of improvement for each bit added to the DAC, as expected.

Figures 3.11 and 3.12 show the demodulated 64-QAM constellation and the distortion PSD at the output of an 8-bit DAC for clipping ratios of 9 dB and 12 dB, respectively. For a 9 dB clipping ratio, the measured EVM is only −33 dB because it is limited by the clipping noise, which is much higher than the quantization noise within the channel. By increasing the clipping ratio to 12 dB, the EVM is significantly improved; as shown in Figure 3.12, the clipping noise becomes negligible and the noise floor is now determined by the DAC quantization noise. We also observe an enhanced constellation, confirming the EVM improvement.

Another impact of the DAC in transmission can be the out-of band emissions generated by the clipping noise, because they can be amplified through the analog transmit path if they are not efficiently filtered. Figure 3.13 depicts the DAC ACLR for WiFi and WiMAX transmitters as a function the clipping ratio and the resolution. Here, the ACLR is defined as the ratio of the power integrated in the adjacent channel (from 10 to 30 MHz offset) to the in-band signal power. As expected, the results are similar for WiFi and WiMAX; and as is the case for EVM, the clipping noise dominates for ratios lower than 10 dB. It is interesting to note that the EVM is more degraded than the ACLR when the clipping noise dominates, that is, at clipping ratios lower than 10 dB, because it is especially concentrated in-band, as we can see from the DAC distortion PSD depicted in Figure 3.11. On the other-hand, for clipping ratios higher than 10 dB,

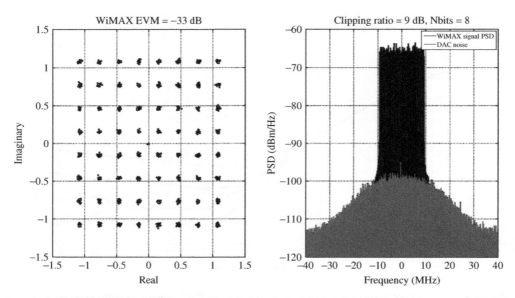

Figure 3.11 WiMAX transmitter 64-QAM constellation and PSD for 8-bit DAC and clip ratio of 9 dB

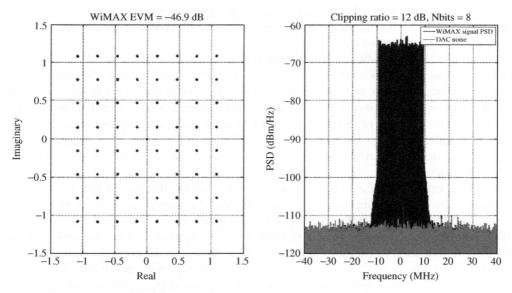

Figure 3.12 WiMAX transmitter 64-QAM constellation and PSD for 8-bit DAC and clip ratio of 12 dB

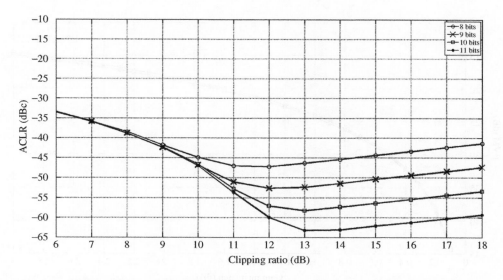

Figure 3.13   Transmitter ACLR at the DAC output vs. DAC clipping ratio and resolution

ACLR and EVM results are well aligned, as the quantization noise (which is spectrally white) dominates over the colored clipping noise.

3.5.3 I and Q Mismatch

We demonstrated in Section 2.7 that the quadrature mixer IQ mismatch generates an image which is the complex conjugate of the signal. The level of this image is defined by the ISR, which is directly related to the gain and phase mismatch values. For an OFDM zero-IF architecture the IQ mismatch creates ICI because the power from subcarrier k will leak into subcarrier $-k$; this ICI can be approximated as Gaussian noise, since the number of modulated subcarriers is large.

In order to separate their contributions, we quantified the impact of the gain and phase imbalances independently.

Figure 3.14 shows the simulated transmitter EVM results as a function of the gain imbalance, in decibels, for WiFi and WiMAX. As expected, the results are identical because the noise level is fixed only by the quadrature mixer ISR, which is determined by the gain mismatch. The impact of 0.3 dB (3.5%) gain mismatch on a 64-QAM constellation is shown in Figure 3.15. The EVM measured on the constellation is −34.9 dB, which matches the approximation of the ISR for small gain imbalance values (Section 2.7):

$$\mathrm{ISR_{dB}} = 20\log_{10}\left(\frac{10^{0.3/20} - 1}{2}\right) = 20\log_{10}\left(\frac{3.5\%}{2}\right) = -35.1\,\mathrm{dB} \tag{3.3}$$

confirming the simulation results.

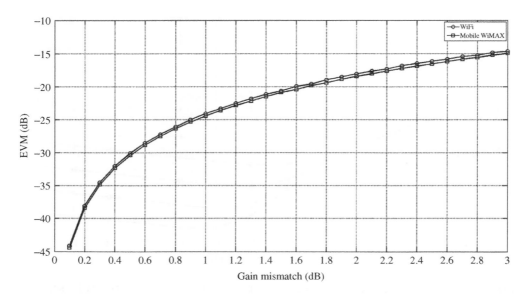

Figure 3.14   WiMAX transmitter EVM as a function of the quadrature gain mismatch

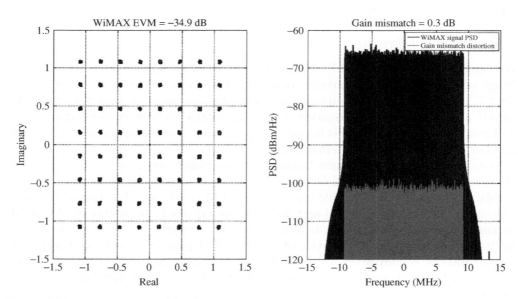

Figure 3.15   WiMAX transmitter 64-QAM constellation and distortion PSD with gain mismatch of 0.3 dB

The complex conjugated image generated by the 0.3 dB gain mismatch can be observed on the right-hand side in Figure 3.15, where the PSD of the WiMAX signal is in black and the IQ mismatch image PSD is superimposed in gray. We see that the image is clearly located within the channel, as expected for a zero-IF transmitter, and its level is 35 dB below the signal level, illustrating the source of the constellation EVM degradation.

Figure 3.16 shows the transmitter EVM as a function of IQ phase mismatch. For comparison we can see that a 2° phase mismatch is equivalent to a 0.3 dB gain error in terms of EVM performance, as confirmed by the ISR computation (Section 2.7):

$$\text{ISR}_{\text{dB}} = 20 \log_{10} \left[\frac{2(\pi/180)}{2} \right] = -35.1 \text{ dB} \tag{3.4}$$

The complex constellation and the PSD of the IQ mismatch image for a 2° phase error are depicted in Figure 3.17. Figure 3.18 shows the simulation results obtained by combining the two mismatches, that is, 0.3 dB for the gain and 2° for the phase. The EVM of −31.8 dB measured on the 64-QAM constellation is in line with the increase of the ISR given by the quadratic sum of the two mismatches (Section 2.7):

$$\text{ISR}_{\text{dB}} = 20 \log_{10} \left\{ \sqrt{ \left(\frac{10^{0.3/20} - 1}{2} \right)^2 + \left[\frac{2(\pi/180)}{2} \right]^2 } \right\} = -32.1 \text{ dB} \tag{3.5}$$

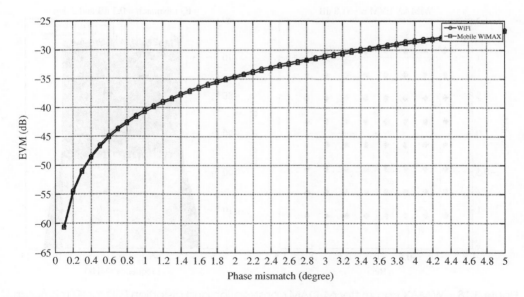

Figure 3.16 WiMAX transmitter EVM as a function of the quadrature phase mismatch

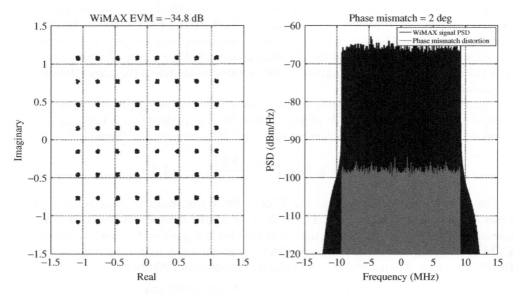

Figure 3.17 WiMAX transmitter 64-QAM constellation and distortion PSD with phase mismatch of 2°

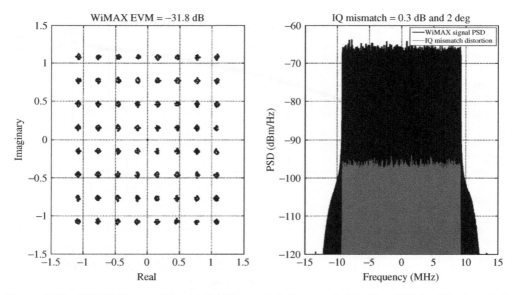

Figure 3.18 WiMAX transmitter 64-QAM constellation and distortion PSD for IQ mismatch of 0.3 dB and 2°

These simulation results demonstrate that IQ phase and gain mismatches have the same impact on the OFDM signal and, therefore, it is very difficult to dissociate them in measurements.

3.5.4 RF Oscillator Phase Noise

We deeply studied in Section 2.3 the impact of the RF oscillator phase noise in communication systems, and in particular those using OFDM technology. We demonstrated that it can be decomposed into two terms: the CPE, rotating all the subcarriers by a common offset; and ICI, generating a Gaussian noise contribution across all the subcarriers. Whereas the CPE can be compensated in reception after the FFT, that is, in the frequency domain, via the de-rotation of the subcarriers performed during channel equalization, the ICI will remain and will limit the system performance. CPE correction will be effective in maintaining performance only if the phase noise bandwidth is smaller than the subcarrier spacing, otherwise the irreducible ICI noise will dominate the result. For the RF oscillator phase noise modeling and generation, we will use the profile depicted in Figure 3.19 because it well represents PLL phase noise profiles that we observe for LOs used in modern transceivers. As a reference phase noise profile, we will consider these values:

- $L_0 = -95\,\text{dBc/Hz}$,
- $L_{\text{floor}} = -150\,\text{dBc/Hz}$,
- $f_{\text{corner}} = 1\,\text{kHz}$,
- $B_{\text{PLL}} = 100\,\text{kHz}$,

giving an integrated phase noise of $0.4°$, that is, $-41.3\,\text{dBc}$.

Figure 3.20 shows the EVM for WiFi and WiMAX transmitters as a function of the PLL phase noise bandwidth varying from $10\,\text{kHz}$ to $1\,\text{MHz}$. All the other PLL phase noise profile parameters are fixed, resulting in an integrated phase noise which varies from around $0.15°$ $(-51.6\,\text{dBc})$ to $1.3°$ $(-33\,\text{dBc})$, as shown in Figure 3.21. We can clearly

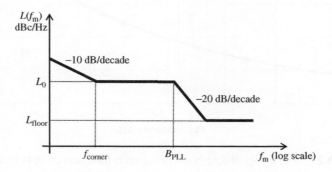

Figure 3.19 PLL double-side band (DSB) phase noise profile in dBc/Hz

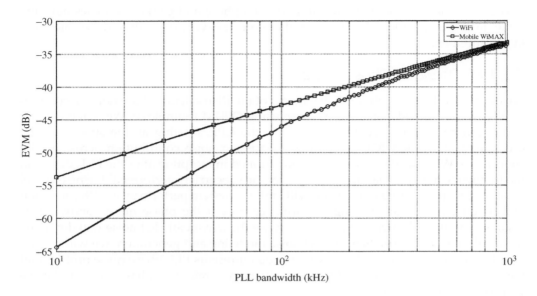

Figure 3.20   WiFi and WiMAX transmitters EVM as a function of the PLL bandwidth

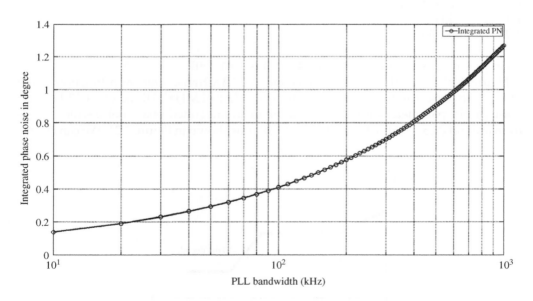

Figure 3.21   PLL integrated phase noise as a function of its bandwidth

see that WiFi presents better performance than WiMAX for small PLL bandwidths, notably until 300 kHz, which is approximately the WiFi subcarrier spacing. This is explained by the fact that, for PLL bandwidths smaller than 300 kHz, CPE dominates in WiFi and can be compensated in reception by the pilot-aided channel estimation which corrects the low-frequency phase noise. However, by choosing a PLL bandwidth which is always equal to or greater than 10 kHz, which is close to the 10.94 kHz WiMAX subcarrier spacing, the ICI generated by the RF oscillator phase noise dominates in the WiMAX transceiver and CPE tracking cannot effectively recover the performance. In this simulation the EVM measured for WiMAX is roughly the ICI given by the integrated phase noise profile reference: −43.1 dBc for PLL bandwidth of 100 kHz, which is a realistic value seen in present-day transceivers.

Figures 3.22 and 3.23 present WiFi and WiMAX demodulated 64-QAM complex constellations and the ICI noise PSD for the reference phase noise profile with 100 kHz PLL bandwidth. Both constellations appear quite clean, demonstrating good EVM but as indicated in the figures, the EVM for WiFi (−46 dB) is 3 dB better than for WiMAX. As explained earlier, CPE can be effectively compensated for WiFi giving a gain of 3 dB because the PLL bandwidth is narrower than the subcarrier spacing, whereas in WiMAX ICI dominates due to the smaller subcarrier spacing. On the PSD we can see that the ICI due to the phase noise, superimposed in gray on the signal PSD, is spread over the entire channel bandwidth.

Figure 3.24 shows simulation results for a configuration where the PLL integrated phase noise remains constant, by scaling the in-band phase noise, while the PLL

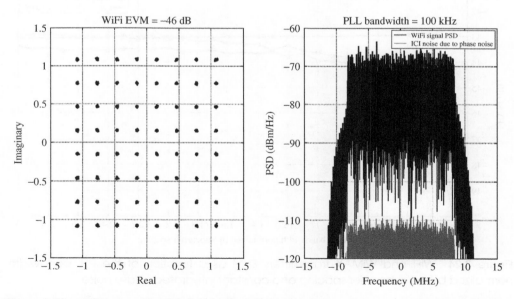

Figure 3.22 WiFi transmitter 64-QAM constellation and ICI with PLL bandwidth of 100 kHz, in-band noise of −95 dBc/Hz

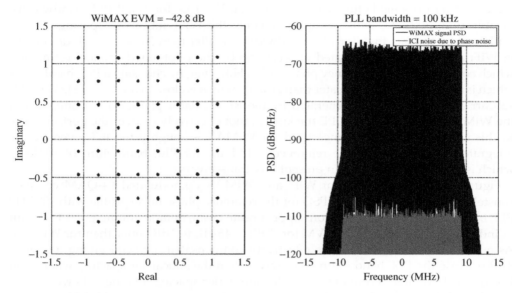

Figure 3.23 WiMAX transmitter 64-QAM constellation and ICI with PLL bandwidth of 100 kHz, in-band noise of −95 dBc/Hz

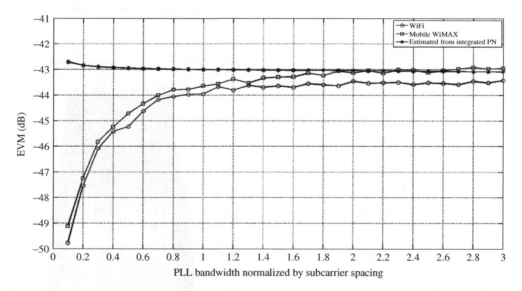

Figure 3.24 WiFi and WiMAX transmitters EVM as a function of the PLL bandwidth normalized by the subcarrier spacing and constant integrated phase noise

bandwidth varies. The aim is to quantify the gain of the CPE compensation. Another important point is to normalize the PLL bandwidth relative to the subcarrier spacing so as to be able to objectively compare WiFi and WiMAX performances. The integrated phase noise is kept equal to −43.1 dBc (the value for the reference PLL noise profile with 100 kHz bandwidth and −95 dBc/Hz in-band noise) while the bandwidth is varied from 1/10 to 3 times the subcarrier spacing. Specifically, the PLL bandwidth varies from 31.25 to 937.5 kHz for WiFi and from 1.094 to 32.82 kHz for WiMAX. We can see that the results are similar for both WiFi and WiMAX and we clearly observe an improvement in the EVM if the PLL bandwidth is smaller than the subcarrier spacing, since CPE dominates over ICI and can be effectively tracked in reception. We estimate approximately 2 dB improvement for a PLL bandwidth which is half of the subcarrier spacing and around 6 dB for a factor of 10. For PLL bandwidths larger than the subcarrier spacing, the EVM tends towards the PLL integrated phase noise value of −41.3 dBc because the ICI noise is the major contributor.

Figure 3.25 is an illustration of the previous discussion using a 16-QAM constellation in order to demonstrate the impact of the CPE on the complex constellation. In this case the PLL bandwidth has been set to 1/10 of the subcarrier spacing and the integrated phase noise is approximately −28 dBc. On the left in Figure 3.25 there is no CPE correction, resulting in a rotation of the constellation points and a measured EVM of −27.7 dB due to the PLL integrated phase noise. On the right in Figure 3.25 the CPE is corrected giving around 7 dB improvement in EVM, and thus a cleaner constellation for the demodulation.

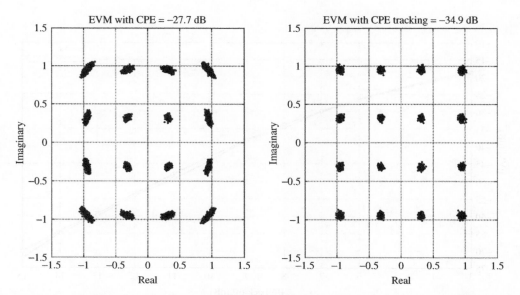

Figure 3.25 WiMAX transmitter 16-QAM constellation with and without CPE tracking (integrated phase noise: −28 dBc)

3.5.5 Power Amplifier Distortion

Due to the intrinsically large PAPR of OFDM signals (Figures 3.7 and 3.8), the performance of an OFDM transceiver is extremely sensitive to the receive and transmit path nonlinearities and the resulting in-band distortions which corrupt the EVM, and out-of-band emissions which generate spectral regrowth. In transmission the principal bottleneck is the PA, which has to be very linear while at the same time providing high output power. Its specifications are either imposed by high-order modulations requiring very low EVM (lower than −30 dB for WiMAX and WiFi 64-QAM) or by the transmission spectral mask for protecting adjacent channels, or both. Contrary to the DAC and ADC clipping which hard-limits the signal above a defined level as simulated in Section 3.5.2, the PA acts as a soft clipping compressing the signal as its level increases. One solution to achieve very linear amplification is to impose a large margin between the PA 1 dB compression point (commonly used to define the distortion limit) and the OFDM signal average power; this is called the back-off. Whereas constant envelope transmitters use high-efficiency switching PAs, OFDM PAs present very poor efficiency due to this back-off requirement. Simulation results plotted in Figure 3.26 show the transmitter EVM as a function of the PA input back-off for both WiFi and WiMAX complex baseband signals. Here, we assumed a Rapp model for the PA (Rapp, 1991) with a smoothness factor of 3, and the 3 dB PAPR of the carrier frequency has not been taken into account. It is important to note that we observe similar performance for WiFi and WiMAX since, as for the ADC and DAC clipping and resolution, the impact depends primarily on the statistical distribution of the signal amplitude, which

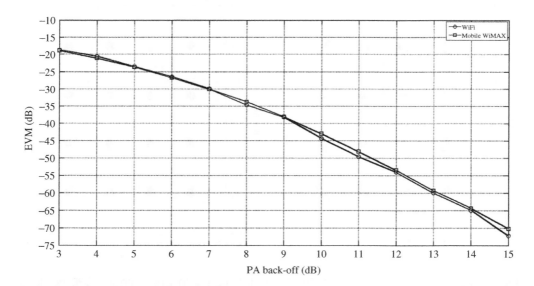

Figure 3.26   TX EVM as a function of the PA input back-off (complex baseband signal)

is similar for both. In order to guarantee an EVM lower than −30 dB as required for the 64-QAM constellation, we can see from the curves that the PA back-off has to be greater than 7 dB. In reality we use larger back-off values in order to take into account the other noise and distortion contributors like IQ mismatch, phase noise, DAC clipping, and resolution previously described and simulated.

Figure 3.27 shows the 20 MHz ACLR estimated for both WiFi and WiMAX signals. Here, ACLR is calculated in the frequency domain, using our spectrum analysis tool, as the ratio between the PA distortions integrated over the adjacent channel bandwidth (20 MHz) and the in-band signal power. In our study case, WiFi presents better ACLR merely because its channel bandwidth is narrower than that of WiMAX, namely 16.25 vs. 18.37 MHz. If we target an ACLR better than −45 dBc, we need 7 dB back-off for WiFi and 8 dB for WiMAX which are again in line with the −30 dB EVM requirements for 64-QAM.

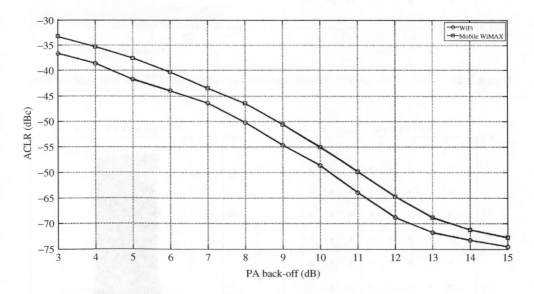

Figure 3.27 TX ACLR at the PA output vs. PA input back-off

Figures 3.28 and 3.29 present the WiMAX 64-QAM constellation and PA distortion PSD simulated for 6 dB and 9 dB PA back-off values, respectively. With these plots we can visually quantify the 12 dB improvement on the constellation for 9 dB back-off versus 6 dB, giving EVMs of −38.2 and −26.6 dB, and explained by the reduction of the in-band distortions caused by the PA compression. In addition we see that the distortions generated out-of-band are significantly reduced by increasing the PA back-off from 6 to 9 dB, which confirms the ACLR improvement observed in the previous simulation results.

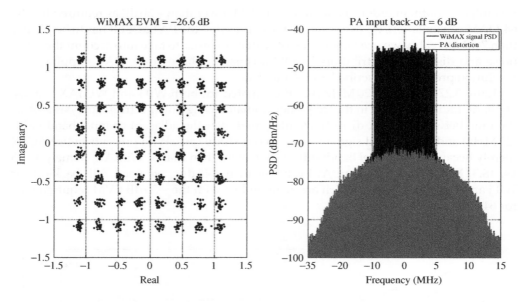

Figure 3.28   WiMAX 64-QAM constellation and PA distortion PSD for back-off of 6 dB

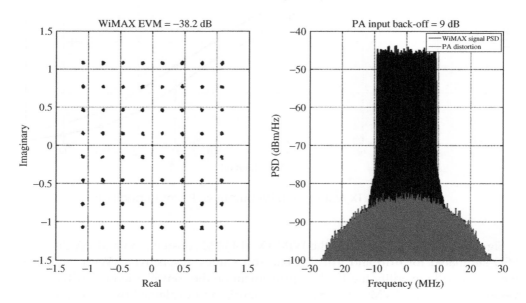

Figure 3.29   WiMAX 64-QAM constellation and PA distortion PSD for back-off of 9 dB

3.5.6 Transmitter Complete Simulation

In this section we will present the simulation results for a WiMAX transmitter taking into account all the RF analog impairments previously simulated. The goal is to study the margin we have for guaranteeing an EVM lower than −30 dB required by the 64-QAM modulation (Table 3.3).The simulation consists of varying the WiMAX transmitter output power from 0 to 25 dBm and measuring the EVM. To allow a detailed evaluation, we added the impairments one by one in order to weigh their impact on the overall transmitter performance. The results are presented in Figure 3.30. The assumed RF analog impairments are

- DAC: 10 bits, clipping ratio 12 dB;
- IQ mismatch 0.1 dB, 1°;
- Phase noise: PLL bandwidth 100 kHz, in-band noise −95 dBc/Hz, giving an integrated phase noise of −43.1 dBc;
- PA 1 dB compression point 30 dBm.

In the first simulation we only took into account the PA and the DAC. We see that for an output power lower than 17 dBm the EVM easily meets the requirement and is limited only by the DAC noise floor below −50 dB; indeed, for these power levels the PA distortion is negligible because the back-off is sufficient. However, for output

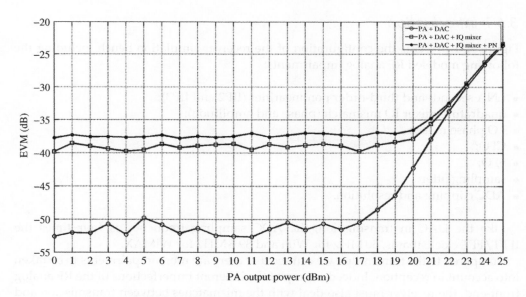

Figure 3.30 Transmitter complete simulation

powers higher than 17 dBm the transmitter EVM begins to degrade due to the PA distortion. In this case an EVM lower than −30 dB is guaranteed if the output power remains below 23 dBm, corresponding to a PA back-off greater than 7 dB.

By adding the mixer IQ mismatch, the EVM floor at low power is severely degraded by more than 10 dB; it is now limited by the ISR of the mixer, which is higher than −40 dB. However, the PA is still the limiting block at high power and an EVM lower than −30 dB is assured only if the output power is restricted to 23 dBm.

Finally, by including the oscillator phase noise, the last simulation result includes the impact of all the RF analog impairments on the WiMAX transmitter performance. We observe that the integrated phase noise contribution is not negligible at low power, increasing the EVM by 1–2 dB, but for meeting the −30 dB EVM specification the PA remains the major factor and always limits the power to 23 dBm.

From these simulations we showed that an EVM lower than −30 dB as specified for 64-QAM modulation is especially limited by the PA distortion if we want to deliver an output power above 23 dBm. On the other hand, for output powers below 20 dBm EVM is mainly limited by the IQ mismatch and the phase noise but remains below −35 dB. Between 20 and 23 dBm, all three of these impairments contribute. The 10-bit DAC resolution and 12 dB clipping ratio have a negligible impact on the EVM.

3.6 Receiver Impairments Simulation

3.6.1 Introduction

Figure 3.31 presents the configuration of the receiver simulation bench including the following modeled RF analog impairments:

- LNA second- and third-order nonlinearities (IIP2 and IIP3);
- quadrature mixer IQ mismatch (phase and amplitude);
- LO phase noise ;
- CFO;
- SFO;
- sampling jitter;
- ADC clipping and resolution.

Like the DAC in transmission, the ADC sampling frequency is four times the IFFT/FFT clock, that is, 80 MHz for WiFi and 89.6 MHz for WiMAX.

Compared to the transmitter, we can see that there are more impairments to be taken into account in reception. Indeed, on top of the inherent imperfections of the RF analog front-end, the receiver must also deal with the mismatches between transmission and reception such as CFO and SFO, which are critical factors in the OFDM demodulation performance.

Like we did for the transmission path, we will study each of these impairments separately in order to isolate their impact on the receiver performance.

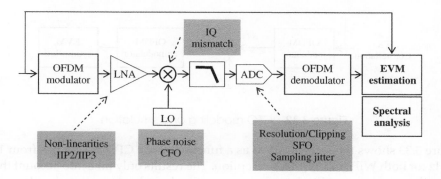

Figure 3.31 RX AFE impairments simulation bench

Regarding the demodulation, perfect OFDM symbol time synchronization is assumed and channel estimation is done using the pilots in order to equalize the FFT output before subcarrier demapping and EVM estimation.

3.6.2 Carrier Frequency Offset

The CFO comes from the difference between the transmitter carrier frequency and the receiver down-conversion frequency arising from two different LOs. Generally, the frequency precision is expressed in ppm and on first consideration would seem small enough to be neglected; but, when it is converted to an absolute frequency, especially in the case of RF communications using gigahertz carrier frequencies, it can be considerable. For example, in the case of WiFi and WiMAX transceivers working in the 2.5 GHz band, 10 ppm precision corresponds to a frequency error of 25 kHz, which is around 2.3 times larger than the subcarrier spacing used in WiMAX!

In addition to the CFO introduced by the transceiver oscillator precision, mobile systems also introduce a speed-dependent shift of the carrier frequency due the Doppler effect. For example, a car moving at 50 km/h will introduce an additional CFO of 116 Hz for a communication link using a 2.5 GHz carrier frequency.

In practice, OFDM systems are more sensitive to CFO than single-carrier systems, especially when a small subcarrier frequency spacing is used, corresponding to long OFDM symbols in the time domain. We have seen in Section 2.5 that CFO in OFDM results in a loss of subcarrier orthogonality and introduces ICI. Actually, the impact of the CFO can be decomposed into two terms:

- A common amplitude attenuation and phase rotation for all the subcarriers, which are generally compensated by the channel estimation and equalization.
- ICI, which limits the receiver EVM performance due to its Gaussian noise properties.

The CFO modeling and its simulation are straightforward, as it consists only of frequency shifting the complex baseband signal between the transmitter and the receiver, as illustrated in Figure 3.32.

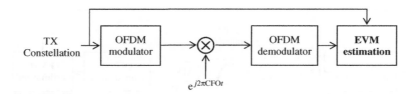

Figure 3.32 CFO modeling and simulation

Figure 3.33 shows the receiver EVM as a function of the CFO as it varies from 10 Hz to 1 kHz for both WiFi and WiMAX reception. The results only take into account the ICI because the common amplitude and phase rotation have been estimated and corrected with the channel estimation and equalization, respectively.

As expected, WiMAX presents worse performance than WiFi because its subcarrier spacing is smaller: 10.94 vs. 312.5 kHz. We can read that a CFO of 100 Hz, that is, only 0.04 ppm at 2.5 GHz, gives an EVM of −35 dB for WiMAX, which is 30 dB higher than for WiFi. Note that the difference between WiMAX and WiFi EVM is constant around 30 dB and both have slopes of 20 dB per decade, which is in line with the theoretical approximation of the normalized ICI noise variance due to CFO (Pollet *et al.*, 1995):

$$\sigma_{\text{ICI}}^2 = \frac{\pi^2}{3} \left(\frac{\text{CFO}}{\Delta f} \right)^2 \tag{3.6}$$

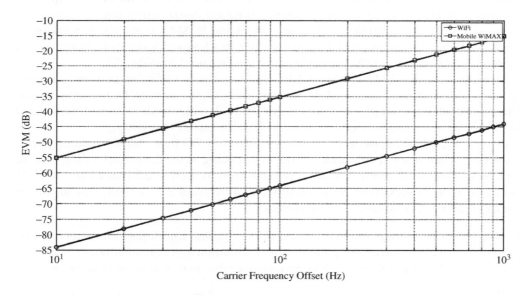

Figure 3.33 WiFi and WiMAX receivers EVM as a function of CFO in hertz

giving in decibels

$$10 \log_{10}(\sigma_{\text{ICI}}^2) \approx 5.2 + 20 \log_{10} \left(\frac{\text{CFO}}{\Delta f} \right) \tag{3.7}$$

where CFO is the carrier frequency offset and Δf is the subcarrier spacing, both in hertz. Equation 3.7 clearly demonstrates that the EVM caused by CFO varies with a slope of 20 dB per decade. The 30 dB performance difference between WiFi and WiMAX for the same CFO can be derived by computing the ratio of the ICI power, or the difference in decibels; given the subcarrier spacings:

$$20 \log_{10} \left(\frac{\Delta f_{\text{WiFi}}}{\Delta f_{\text{WiMAX}}} \right) = 20 \log_{10} \left(\frac{312.5 \, \text{kHz}}{10.94 \, \text{kHz}} \right) = 29.1 \, \text{dB} \tag{3.8}$$

Figures 3.34 and 3.35 depict 64-QAM constellations for both WiFi and WiMAX with CFO equal to 1 kHz and 100 Hz, respectively. We clearly see the greater robustness of WiFi with clearly distinguishable constellations for 1 kHz and 100 Hz CFOs which exhibit EVMs of -44 dB and -64 dB, respectively. On the other hand, the WiMAX constellation with a 1 kHz CFO is very noisy and will be very difficult to demodulate with a measured EVM of only -15 dB versus the 64-QAM constellation requirement of less than -20 dB. By reducing the CFO to 100 Hz the EVM is improved by 20 dB, as predicted by Equation 3.7; as a result the constellation becomes much cleaner.

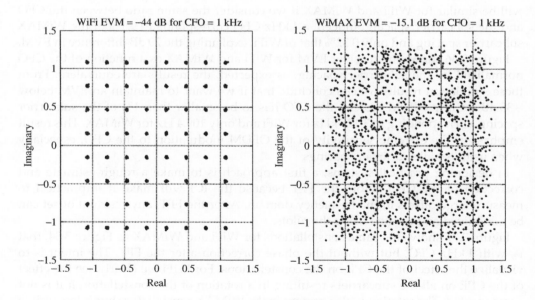

Figure 3.34 RX WiFi and RX WiMAX 64-QAM with CFO = 1 kHz

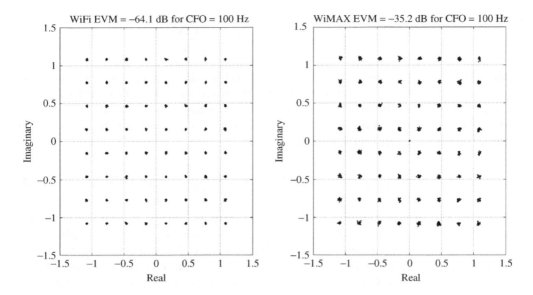

Figure 3.35   WiFi and WiMAX receivers 64-QAM constellation with CFO = 100 Hz

It is interesting to note from Equation 3.7 that the ICI power generated by the CFO will be similar for WiFi and WiMAX if we consider the same ratio between the CFO and the subcarrier spacing. Indeed, a 1 kHz CFO corresponds to 9.1% of the WiMAX subcarrier spacing and only 0.32% that of WiFi, explaining the 29 dB difference in EVM.

Figure 3.36 shows the receiver EVM for WiFi and WiMAX as a function of the CFO normalized by the subcarrier spacing; as expected, the results are equivalent. From these simulation results we can conclude that if we want to maintain an EVM below −35 dB for any OFDM system, the CFO has to be smaller than 1% of the subcarrier spacing, corresponding to 3.125 kHz for WiFi and only 109.4 Hz for WiMAX. This result emphasizes the extreme sensitivity of the OFDM modulation to the CFO, especially when using small subcarrier spacings.

In designing practical systems, a first approach is to make a rough estimate and correction of the CFO before the FFT because the ICI will make it impractical to measure large offsets in the frequency domain. After the FFT, any residual offset can be minimized and tracked using the pilots.

Figure 3.37 shows the same constellations for WiFi and WiMAX as Figure 3.34, that is, with 1 kHz CFO, but without the phase correction after the FFT. The intent is to visualize the effect of the CPE on the constellations. For WiFi we clearly see the effect of the CPE on all the subcarriers resulting in a rotation of the constellation if it is not compensated. The rotation is also present in the WiMAX constellation but is less visible because the ICI is much more than in WiFi.

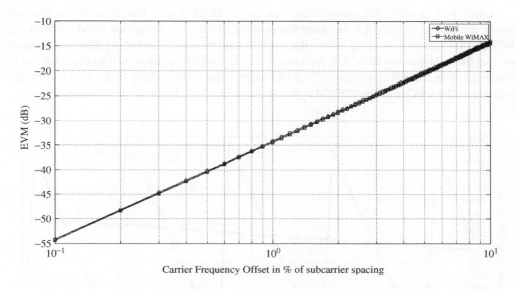

Figure 3.36   RX EVM as a function of CFO normalized by the subcarrier spacing

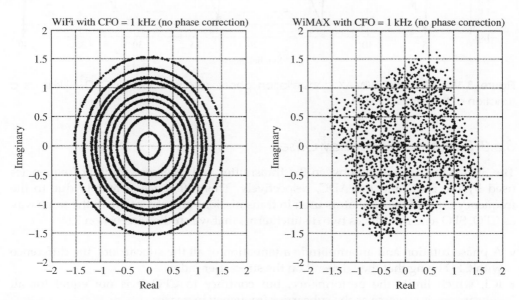

Figure 3.37   RX WiFi and RX WiMAX 64-QAM constellations for CFO = 1 kHz without phase correction

Figure 3.38 depicts the phase shift due to 1 kHz CFO as a function of the symbol index for WiFi and WiMAX. Because the symbol duration is larger for WiMAX than for WiFi, 114.3 versus 4 μs, the phase rotation between two consecutive symbols is around of 28.5 times larger for WiMAX. As a result, the complete rotation of the constellation is obtained only after 9 symbols for WiMAX compared with 250 for WiFi; this is directly related to the CFO and the OFDM symbol duration: $1/(\text{CFO} \times \text{symbol duration})$.

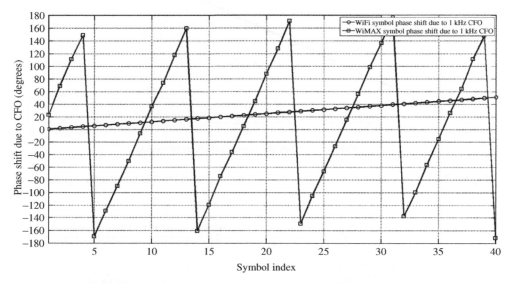

Figure 3.38 WiFi and WiMAX constellation phase rotation due to CFO = 1 kHz as a function of the symbol index

3.6.3 Sampling Frequency Offset

The SFO is the mismatch between the transmitter and receiver sampling frequencies used by the DAC and the ADC, respectively. This difference is merely due to the frequency error of the oscillators used in transmissions and reception. In the same way as CFO, SFO also introduces two distinct terms that we studied in Section 2.6:

- A phase rotation and an amplitude attenuation of all the subcarriers, the difference with CFO being that they depend on the subcarrier index.
- ICI, which limits the performance, but contrary to CFO it is not equal for all subcarriers; it is worse as the subcarrier frequency increases.

Similar to CFO, the first term affecting the amplitude and the phase of the subcarriers is compensated and tracked with the pilots; consequently, we will only focus on the performance degradation introduced by the ICI.

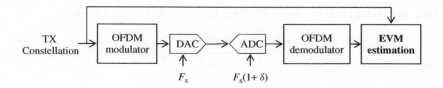

Figure 3.39   SFO modeling and simulation

The modeling and simulation of the SFO is depicted in Figure 3.39. From a simulation point of view the OFDM signal generated by the modulator is not simply resampled in reception but interpolated on a new time vector, compressed or expanded depending on the sign of the sampling frequency error.

Figure 3.40 shows the EVM for WiFi and WIMAX receivers as a function of the SFO as it varies from 1 to 100 ppm. As for CFO, WiMAX is much more sensitive to SFO than WiFi is because of its smaller subcarrier spacing. Indeed, for the same SFO, WiFi always presents an EVM 30 dB better than WiMAX, as was observed in the CFO simulation results. For example, a 10 ppm SFO results in a −42 dB EVM for WiMAX as opposed to −72 dB for WiFi. In addition, the EVM varies with a slope of 20 dB per decade versus the SFO, as for the CFO results. Actually, all these common points can be explained because the normalized ICI noise variance introduced by SFO is similar to CFO, except

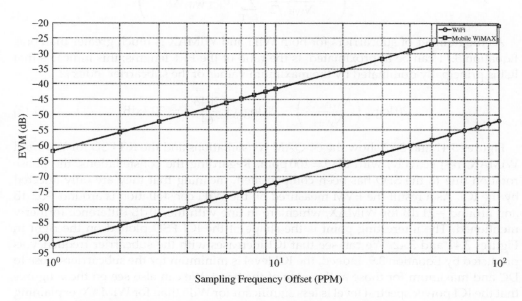

Figure 3.40   RX EVM as a function of the SFO in parts per million

for the fact that it depends on the subcarrier index (Pollet *et al.*, 1994):

$$\sigma_{ICI}^2(k) = \frac{\pi^2}{3}\left(k\frac{SFO}{F_s}\right)^2 = \frac{\pi^2}{3}(k\delta)^2 \tag{3.9}$$

giving in decibels

$$10\log_{10}(\sigma_{ICI}^2) \approx 5.2 + 20\log_{10}(k) + 20\log_{10}(\delta) \tag{3.10}$$

in which SFO is the sampling frequency offset in hertz, F_s is the sampling frequency in hertz, k is the subcarrier index, and δ is the relative SFO. The ICI noise power increases with the subcarrier index because the frequency error relative to the ideal subcarrier position cumulates, as opposed to the effect of CFO, which is independent of subcarrier position. Equation 3.10 clearly shows the 20 dB per decade SFO dependency of the ICI explaining the simulation results of Figure 3.40. The 30 dB EVM difference between WiFi and WiMAX performances for the same normalized SFO value can be determined from the average ICI power ratio expressed in decibels:

$$10\log_{10}\left(\frac{\dfrac{2}{N_{WiMAX}/2}\displaystyle\sum_{k=1}^{N_{WiMAX}/2}\sigma_{ICI\_WiMAX}^2(k)}{\dfrac{2}{N_{WiFi}/2}\displaystyle\sum_{k=1}^{N_{WiFi}/2}\sigma_{ICI\_WiFi}^2(k)}\right) \tag{3.11}$$

As the number of subcarriers for both WiFi and WiMAX is much greater than one, Equation 3.11 can be approximated as the ratio of the FFT lengths; this indicates that ICI is principally dominated by the maximum values of the subcarrier index k:

$$20\log_{10}\left(\frac{N_{WiMAX}}{N_{WiFi}}\right) = 20\log_{10}\left(\frac{2048}{64}\right) = 30.1\,dB \tag{3.12}$$

Figures 3.41 and 3.42 show the 64-QAM constellations and the ICI PSD for WiFi and WiMAX, respectively, with SFO = 100 ppm. Regarding the constellations, the phase rotation due to the SFO has been compensated, meaning that they are only affected by the ICI. As a result the EVM measured on the WiFi constellation is around $-52\,dB$ and around $-21\,dB$ for WiMAX, which are in line with the 30 dB difference formerly mentioned. The interesting point is the shape of the ICI PSD plotted on the right in Figures 3.41 and 3.42. We can see that ICI increases with the subcarrier frequency as predicted by Equation 3.9. Indeed, the ICI level is minimum for the subcarriers close to DC and maximum for those at the edge of the band. We can also see on these figures that the ICI power spectral level is less significant for WiFi than for WiMAX, explaining the constellation quality difference.

Figure 3.43 shows the EVM per subcarrier frequency, *and not per subcarrier index*, for both WiFi and WiMAX, again for a 100 ppm sampling frequency precision. We can see

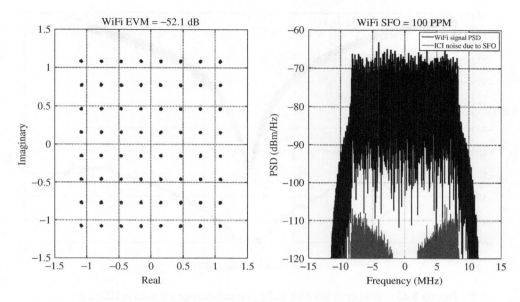

Figure 3.41 WiFi 64-QAM constellation and ICI noise PSD with SFO = 100 ppm

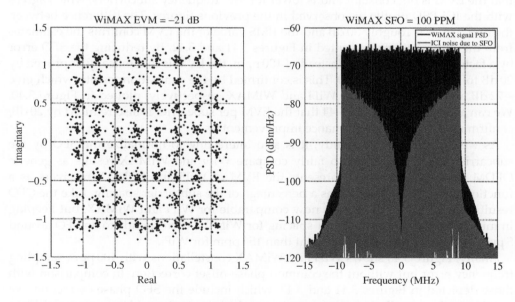

Figure 3.42 WiMAX 64-QAM constellation and ICI noise PSD with SFO = 100 ppm

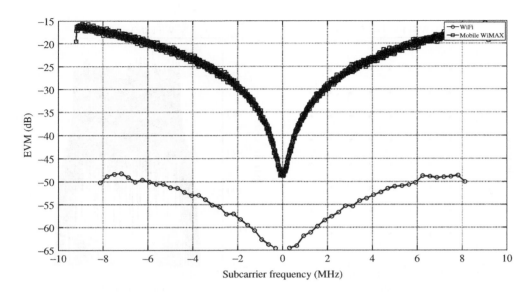

Figure 3.43   WiFi and WiMAX EVM per subcarrier for SFO $= 100$ ppm

that the EVM is not constant and is lower for low-frequency subcarriers, which agrees with the ICI PSD profile we observed in the previous figures. The difference between the two curves is roughly 30 dB and the RMS value of the EVM confirms the estimates from the constellations presented in Figures 3.41 and 3.42. By reducing the SFO error by a factor 10, that is, 10 ppm instead of 100 ppm, the EVM is theoretically improved by 20 dB following Equation 3.10. This is confirmed by the simulation results, which give -72 dB and -42 dB EVM for WiFi and WiMAX, respectively, plotted in Figure 3.40. We can also notice in Figure 3.44 that the EVM per subcarrier is also reduced by 20 dB, confirming the overall performance improvement.

As we did for the CFO study, it is also interesting to normalize the SFO by the subcarrier spacing in order to fairly compare WiFi and WiMAX results as generic OFDM systems. Figure 3.45 presents the EVM of WiFi and WiMAX receivers as a function of the SFO expressed as a percentage of the subcarrier spacing. Like the CFO results, the impact of the SFO is now comparable for WiFi and WiMAX, but keeping in mind that 1% of the subcarrier spacing for WiMAX corresponds to an SFO around 5 ppm, which is much more stringent than 156 ppm for WiFi.

Finally, Figure 3.46 shows WiFi and WiMAX constellations with 100 ppm sampling frequency error but without the common phase-offset correction. In comparison with those depicted in Figures 3.41 and 3.42, which include the SFO phase correction, we see that both constellations appear noisier, which is confirmed by a measured EVM degradation of 9 dB. Clearly, the WiMAX constellation is more affected by ICI due to the 100 ppm SFO, representing 20% of the subcarrier spacing for WiMAX and only 0.6% for WiFi.

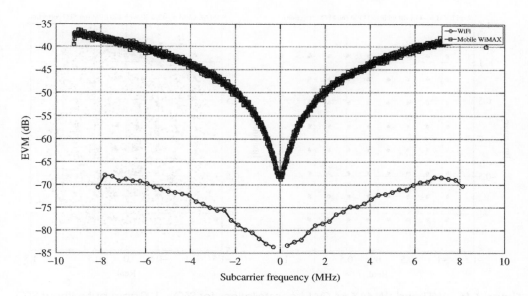

Figure 3.44   WiFi and WiMAX EVM per subcarrier for SFO = 10 ppm

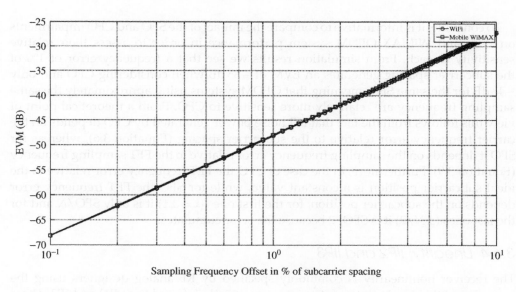

Figure 3.45   RX EVM as a function of SFO normalized by the subcarrier spacing

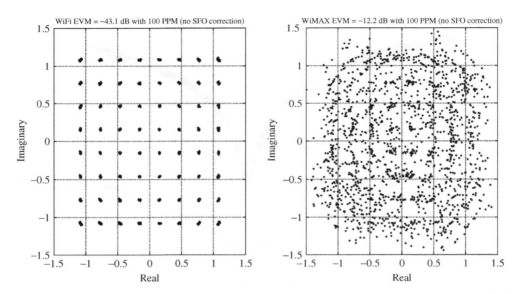

Figure 3.46   WiFi and WiMAX 64-QAM constellations for SFO = 100 ppm but without SFO correction

To conclude, it is informative to compare the impact of the SFO and CFO impairments on WiFi and WiMAX OFDM system performance in order to assess their relative sensitivity to each. From simulation results we see that a frequency error of 1% of the subcarrier spacing generates an EVM of −35 dB when considering CFO and only −48 dB for the same SFO, meaning that OFDM systems using approximately the same sampling frequency are relatively more sensitive to CFO. From a theoretical point of view this follows from the fact that ICI noise power generated by CFO depends on the carrier frequency error relative to the subcarrier spacing (Equation 3.6), whereas for SFO it depends on the sampling frequency error relative to the FFT sampling frequency (Equation 3.9). Furthermore, in the case of CFO the FFT frequency error relative to the ideal subcarrier position is a constant value, while for SFO the FFT frequency error depends on the subcarrier position; for the first ones ($k = \pm 1$) it is only SFO/N and for the last ones ($\pm N/2$) it is SFO/2 because the error cumulates.

3.6.4 Linearity: IIP2 and IIP3

The receiver nonlinearity is commonly specified by RF analog designers using the second- and third-order intercept points, respectively referred to as IP2 and IP3. These nonlinearities can distort the signal or alias out-of-band blockers into the channel, and hence degrade the receiver performance. For a detailed description of the receiver IP2 and IP3, please refer to Section 2.9. Because the impact of IP2 and IP3 is similar for WiFi and WiMAX, as with PA distortion, in this section we will especially focus on the WiMAX receiver but similar results can be developed for WiFi.

3.6.4.1 Impact of Receiver IP3

In the case of an OFDM signal, the receiver third-order nonlinearity generates in-band distortion called composite triple beats (CTBs), which is a consequence of subcarrier intermodulation (Section 2.9.4). The in-band level of the CTB, and thus the EVM, directly depends on the IP3 value and the signal power. We demonstrated that the EVM due only to the CTB can be approximated as

$$\text{EVM}_{\text{dB}} \approx 2(P_{\text{in}} - \text{IP3}) + 1.2 \tag{3.13}$$

Consequently, for a required EVM, the receiver IP3 has to be specified relative to the maximum power that the receiver has to handle. Figure 3.47 shows the EVM simulation results for two different values of IIP3, 0 and 10 dBm, which are fairly representative for modern on-chip receivers. The plots represent the EVM as a function of the receiver input power swept from −90 to 0 dBm. The noise figure of the receiver has been fixed to 5 dB, giving an input-referred thermal noise power of −96 dBm for a 20 MHz channel bandwidth, which explains the EVM close to −6 dB for −90 dBm input power. As the receiver input power increases, we see that the EVM linearly improves decibel per decibel as expected, but from approximately −50 dBm the EVM is limited to approximately −40 dB by the noise floor imposed by other receiver impairments (phase noise, IQ mismatch, etc.).

For IIP3 = 0 dBm we observe a degradation of the EVM starting from −25 dBm input power, whereas it degrades from −15 dBm with IIP3 = 10 dBm. Additionally, as the

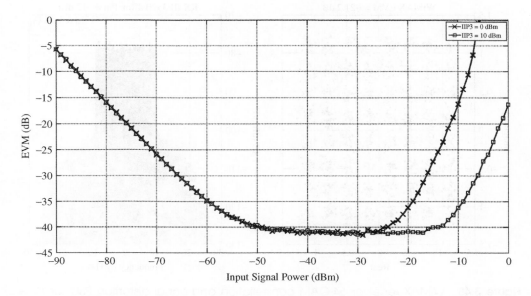

Figure 3.47 RX EVM as a function of the input power for IIP3 = 0 and 10 dBm

signal power continues to increase, the third-order distortions dominate and become the major EVM contributor. Given the minimum SNR requirement for 64-QAM demodulation of 20 dB (Table 3.3), these simulation results demonstrate that performance is only guaranteed if the receiver maximum input power is lower than −12 dBm and −2 dBm for IIP3 = 0 dBm and 10 dBm, respectively.

It is also interesting to remark from these two curves that increasing the IIP3 by 1 dB improves the EVM by 2 dB for the same input power, as predicted by the theory and shown in Equation 3.13. But this formula also shows that, for the same IP3, increasing the input power by 1 dB degrades the EVM by 2 dB.

Figures 3.48 and 3.49 show the 64-QAM constellations and the distortion PSD only due to third-order distortion for −12 dBm input power and receiver IIP3 = 0 dBm and IIP3 = 10 dBm, respectively. For IIP3 = 0 dBm, the 64-QAM constellation appears quite noisy. This is confirmed by the measured EVM around −21 dB, which is at the limit of the −20 dB required for the demodulation of a 64-QAM constellation (Table 3.3). Furthermore, we can see on the right-hand side of the figure the distortion created by the receiver IP3 in the frequency domain, highlighting the in-band CTB limiting the EVM to around −20 dB. With IIP3 = 10 dBm, that is, 10 dB better than the previous simulation, we can see on the spectrum that the distortion due to CTB is reduced and consequently the 64-QAM constellation becomes much cleaner. For this case the measured EVM is around −42 dB, which is 20 dB better than for IIP3 = 0 dBm, as expected by the theory, that is, two times the IP3 gradient.

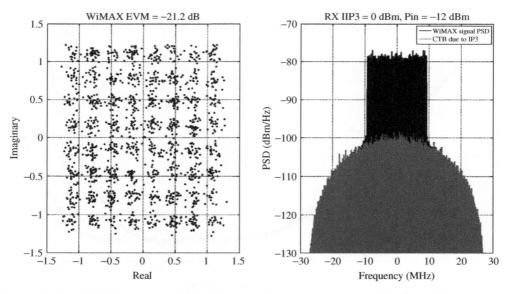

Figure 3.48 WiMAX receiver 64-QAM constellation and signal distortion PSD for P_{in} = −12 dBm and IIP3 = 0 dBm

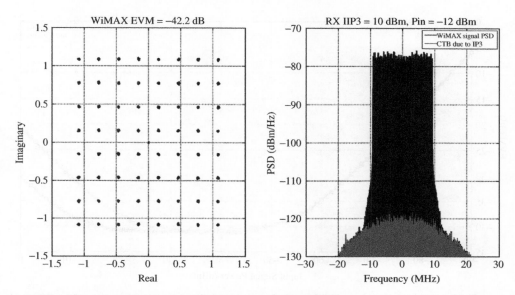

Figure 3.49 WiMAX receiver 64-QAM constellation and signal distortion PSD for $P_{in} = -12\,\text{dBm}$ and $IIP3 = 10\,\text{dBm}$

3.6.4.2 Impact of Receiver IP2

For the simulations concerning the IIP2 impact on the receiver performance, we used the same parameters as those for IIP3: a receiver noise figure of 5 dB and a noise floor around −40 dBc. We can see in Figure 3.50 that the EVM results as a function of IIP2 are analogous to those obtained for IIP3 at low input signal powers.

For the following simulations, we assume two realistic values seen in current CMOS receivers: 20 and 30 dBm. For IIP2 = 20 dBm we observe a degradation of the EVM starting from −30 dBm input signal power, as opposed to −20 dBm for IIP2 = 30 dBm. For 0 dBm input signal power, the EVM obtained with IIP2 = 20 dBm is very close to the −20 dB requirement, whereas we have 10 dB margin with IIP2 = 30 dBm. We can observe on the figure that the EVM degradation linearly increases decibel per decibel with the input signal power because the IMD level due to the second-order nonlinearity is proportional to the square of the signal power; it is expressed in decibels as follows:

$$IMD_{dB} = 2P_{in} - IP2 \tag{3.14}$$

giving an approximation of the EVM due to the second-order distortion:

$$EVM_{dB} = IMD_{dB} - P_{in} = P_{in} - IP2 \tag{3.15}$$

which is in line with the simulation results because most of the IMD2 energy is localized into the channel bandwidth. As an example, if we consider a 0 dBm input signal power

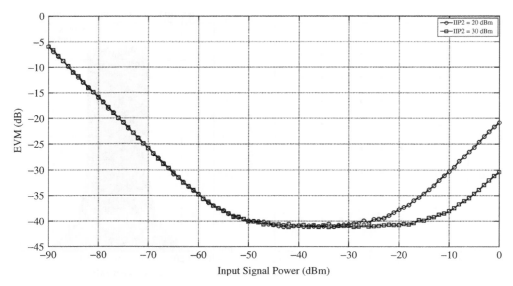

Figure 3.50 WiMAX receiver EVM as a function of the input power for IIP2 = 20 and 30 dBm

with IIP2 = 20 dBm, from Equation 3.15 we find an EVM of −20 dB, which matches quite well with the simulation result plotted in Figure 3.50.

Figures 3.51 and 3.52 illustrate the previous simulation results and analysis by showing the 64-QAM constellations and the distortion PSD, only due to IP2, for 0 dBm input power and receiver IIP2 = 20 and 30 dBm. The EVM levels, and hence the constellation distinguishability, are directly related to the in-band second-order distortion, which is reduced by increasing the IIP2.

3.6.4.3 Impact of Receiver IP2 and IP3

The IP2/3 simulation is a combination of the two previous ones because it takes into account concurrently the receiver IP2 and IP3 in order to study the impact of the overall receiver nonlinearity on the system performance. We assume IIP2 = 30 dBm and IIP3 = 10 dBm, which are realistic figures for today's on-chip receiver. The simulation results are plotted in Figure 3.53. Although IP2 and IP3 start to degrade the EVM around the same input signal power, between −20 and −15 dBm, it is primarily the IP3 which limits the EVM for high input signal power because the CTB, that is, third-order distortion, rises two times faster, in decibels, than the second-order distortion generated by IP2. This result is in line with Equations 3.13 and 3.15, which are the theoretical EVM estimations for IP3 and IP2, respectively.

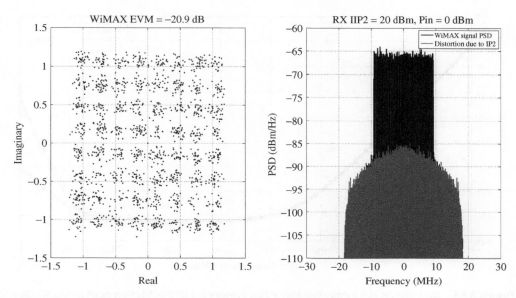

Figure 3.51 WiMAX receiver 64-QAM constellation and signal distortion PSD for $P_{in} = 0$ dBm and IIP2 = 20 dBm

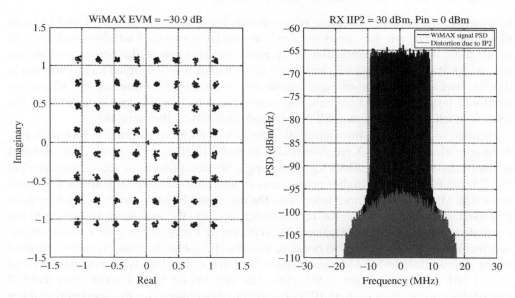

Figure 3.52 WiMAX receiver 64-QAM constellation and signal distortion PSD for $P_{in} = 0$ dBm and IIP2 = 30 dBm

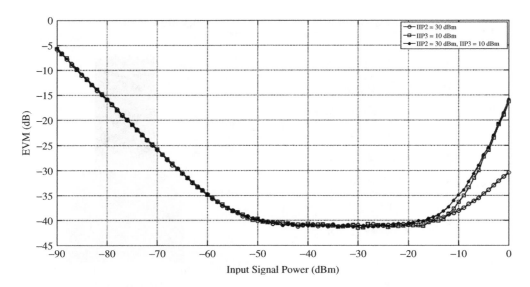

Figure 3.53   RX EVM as a function of the input power for IIP2 = 30 dBm and IIP3 = 10 dBm

3.6.4.4 Linearity versus Out-of-Band Interferers

In practice, the receiver linearity requirements are usually imposed by out-of-band blockers. Indeed, wireless receivers have to coexist with other wireless systems and hence must be able to demodulate very weak signals even in the presence of strong interferers at their input which can be tens of decibels above the signal level. Whereas in digital baseband these out-of-band interferers are removed by filtering, the RF analog front-end (especially the LNA and the mixer) has a bandwidth much wider than the channel itself in order to cover all the channels defined by the standard.

Figures 3.54 and 3.55 show simulation results of an intermodulation test with two different values of WiMAX receiver IIP3: +10 and −10 dBm, respectively. We assumed at the receiver input a WiMAX signal having −60 dBm power, an adjacent channel at 25 MHz offset with 20 dB higher power, and an alternate channel 30 dB above the signal level at 50 MHz offset. Without interferers the measured EVM is around −35 dB because it is fixed by the phase noise, the IQ mismatch, and the thermal noise according to the simulation result plotted in Figure 3.53. With an IIP3 of +10 dBm, we can see on the spectrum that the interferers do not leak into the channel of interest. This is confirmed by the demodulated 64-QAM constellation, which is quite clean with a measured EVM of −34.3 dB matching the predicted value due to the noise floor. On the other hand, if the receiver is specified with an IIP3 of −10 dBm, we observe on the spectrum that the third-order distortion created by the interferers becomes higher than the channel noise floor and degrades the system performance. The direct impact is a noisier 64-QAM

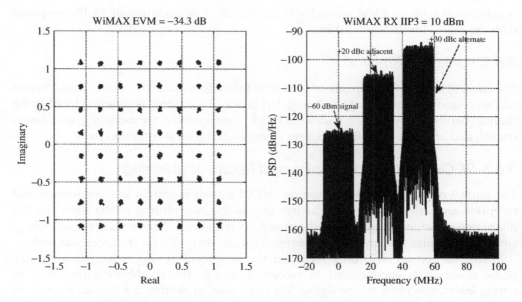

Figure 3.54 WiMAX receiver intermodulation test with IIP3 = +10 dBm

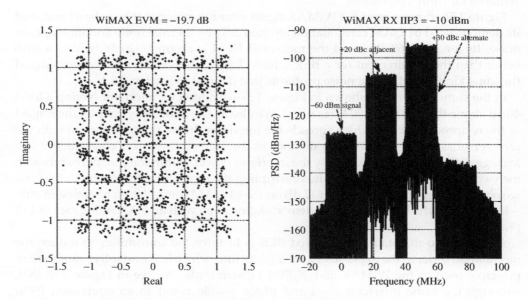

Figure 3.55 WiMAX receiver intermodulation test with IIP3 = −10 dBm

constellation with an EVM around -19 dB, that is, a degradation of 15 dB compared with the case with IIP3 $= +10$ dBm.

3.6.5 I and Q Mismatch

The effect of mixer I and Q mismatch on OFDM signals is similar in zero-IF transmission and reception, in that a complex conjugated image of the signal is generated within the channel bandwidth which degrades the EVM. Consequently, the modeling, simulation results, and analysis of Section 3.5.3 are equally valid for the reception path.

3.6.6 RF Oscillator Phase Noise and Reciprocal Mixing

The impact of oscillator phase noise on OFDM signals is similar in transmission and reception and affects all the subcarriers due to the convolution of the signal by the phase noise profile introduced by the mixer. As with I and Q mismatch, the modeling, simulation results, and analysis of Section 3.5.4 are also valid for the reception path.

In this section we will only focus on the reciprocal mixing, which is a specific phase noise issue encountered in reception due to the neighboring channels, or strong interferers close to the signal. We have seen in Section 2.3 that all interferers present at the input of the mixer will also be convolved by the oscillator phase noise, and consequently may reinject phase noise into the signal bandwidth despite being centered far from the channel.

Figure 3.56 shows a received WiMAX signal after the mixer, that is, in baseband, and its demodulated 64-QAM constellation affected only by thermal noise and mixer phase noise. Its level is -60 dBm and the measured EVM is around -28 dB, which is 8 dB better than the requirement for a BER $= 10^{-6}$ (Table 3.3). For this case we considered the same kind of PLL phase noise profile as that defined in Section 3.5.4.

In the simulation result shown in Figure 3.57 we added an interferer having a level 40 dB above the signal at a 30 MHz offset. We clearly see in the spectral plot the impact of the reciprocal mixing, which spreads the interferer due to the convolution with the receiver phase noise in the same manner as the signal. The direct consequence is a leakage of the phase noise scaled by the interferer power within the channel bandwidth increasing the in-band noise and thus corrupting the receiver performance. In this case we observe a degradation of about 12 dB on the 64-QAM constellation resulting in only -16 dB EVM, which is the limit for demodulation even with an FEC coding rate of $1/2$ (Table 3.3).

A solution to maintain the required BER is to force the transmitter to reduce the modulation order at the expense of a lower data rate, which is the adaptive modulation principle used in the WiMAX and 3GPP-LTE standards. We see in Figure 3.58 that, although the same interferer level and phase profile result in an equivalent EVM using a 16-QAM, the reduced-order constellation can still be distinguished, allowing better demodulation.

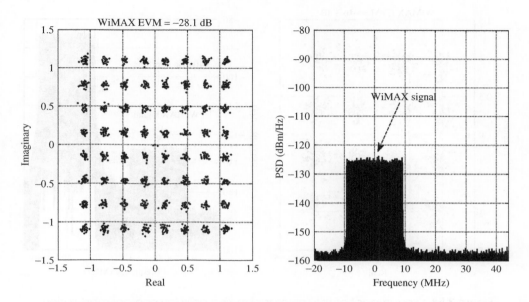

Figure 3.56 Original WIMAX receiver 64-QAM constellation without interferer

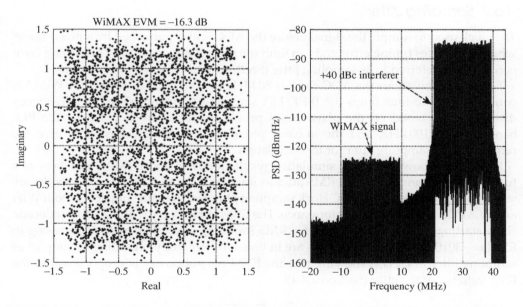

Figure 3.57 WIMAX receiver 64-QAM EVM degradation due to the reciprocal mixing

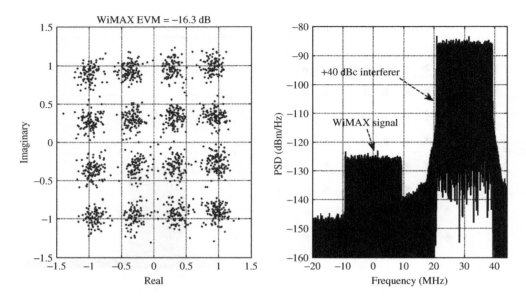

Figure 3.58   Less impact of the reciprocal mixing on a lower order modulation

3.6.7 Sampling Jitter

The clock used to sample the signal before the ADC is also perturbed by a phase noise, which introduces timing error and can limit the system performance. This timing error is commonly referred to as sampling jitter (Section 2.4).

For these simulations the ADC clock is 80 MHz for WiFi and 89.6 MHz for WiMAX, corresponding to four times the IFFT/FFT sampling frequency. The jitter frequency distribution is based on the phase noise profile described in Figure 3.19, the PLL bandwidth is 100 kHz, and the in-band phase noise level is adjusted to give the required RMS sampling jitter for the simulation.

Figure 3.59 shows the EVM simulation results for WiFi and WiMAX receivers as a function of the sampling clock RMS jitter as it varies from 10 ps to 1 ns. The results are similar, indicating that the ADC clock sampling jitter has the same impact on both WiFi and WiMAX demodulation performance. The EVM curves vary with a 20 dB per decade slope starting from EVM = −70 dB for RMS sampling jitter of 10 ps and increasing to EVM = −30 dB at 1 ns. These results are in line with our study regarding the impact of sampling jitter in OFDM showing that the EVM per subcarrier is proportional to the RMS value σ_j of the jitter (Section 2.4.4):

$$\text{EVM}(k) = \sqrt{4\pi^2(k\Delta f)^2\sigma_j^2} \tag{3.16}$$

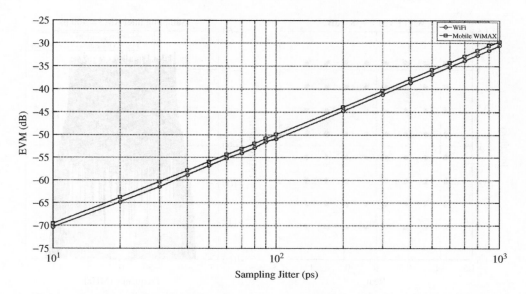

Figure 3.59   WiFi and WiMAX receivers' EVM as a function of the ADC sampling jitter

giving the 20 dB per decade in log scale:

$$\text{EVM}_{\text{dB}}(k) \approx 16 + 20\log_{10}(k\Delta f) + 20\log_{10}(\sigma_{\text{j}}) \tag{3.17}$$

where k is the subcarrier index, Δf is the subcarrier spacing, and σ_{j} is the RMS of the sampling jitter.

Equation 3.16 also demonstrates that the EVM introduced by the sampling jitter is mainly dominated by the high-frequency subcarriers, explaining the equivalent results between WiFi and WiMAX, which have roughly the same bandwidth. Indeed, the highest subcarrier frequency for WiFi is 8.125 MHz ($52/2 \times 312.5$ kHz) and for WiMAX is 9.1875 MHz ($1680/2 \times 10.9375$ kHz). Actually, if we precisely compare the results we see a small performance difference of around 1 dB in favor of WiFi because the bandwidth is slightly smaller than WiMAX. This 1 dB difference can be predicted from the ratio of the average EVM expressed in decibels as follows:

$$10\log_{10}\left(\frac{\dfrac{2}{N_{\text{WiMAX}}/2}\displaystyle\sum_{k=1}^{N_{\text{WiMAX}}/2}\text{EVM}_{\text{WiMAX}}^2(k)}{\dfrac{2}{N_{\text{WiFi}}/2}\displaystyle\sum_{k=1}^{N_{\text{WiFi}}/2}\text{EVM}_{\text{WiFi}}^2(k)}\right) \tag{3.18}$$

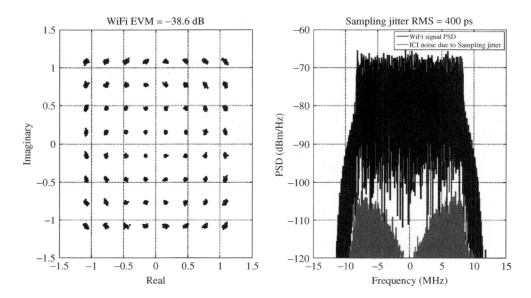

Figure 3.60   WiFi 64-QAM constellation and ICI noise with sampling jitter of 400 ps

which can be approximated as the ratio of the bandwidth if the number of subcarriers is much greater than one:

$$20 \log_{10} \left(\frac{N_{\text{WiMAX}} \Delta f_{\text{WiMAX}}}{N_{\text{WiFi}} \Delta f_{\text{WiFi}}} \right) = 20 \log_{10} \left(\frac{1680 \times 10.9375 \text{ kHz}}{52 \times 312.5 \text{ kHz}} \right) \approx 1 \text{ dB} \qquad (3.19)$$

Figures 3.60 and 3.61 show the 64-QAM constellation and the ICI noise spectrum for WiFi and WiMAX baseband signals, respectively, sampled with an RMS sampling jitter of 400 ps. The constellations are nearly identical with an EVM around −38 dB, confirming the previous simulation results. The interesting point is the ICI noise PSD distribution generated by the sampling jitter; we clearly see on the spectral plots that it is frequency dependent and maximum at the edges of the signal PSD, as predicted by Equations 3.16 and 3.17.

The last simulation results plotted in Figure 3.62 show the EVM per subcarrier for WiFi and WiMAX signals impaired by the same 400 ps sampling jitter. As expected, the plots are identical and match quite well with Equation 3.16 predicting a degradation of the EVM as the subcarrier frequency increases. In addition, these EVM per subcarrier plots confirm the shape of the ICI noise power spectrum of Figures 3.60 and 3.61.

3.6.8 ADC Clipping and Resolution

The impact of the ADC clipping and resolution on the OFDM signal in reception is similar to the one studied for the DAC in transmission. Consequently, the modeling,

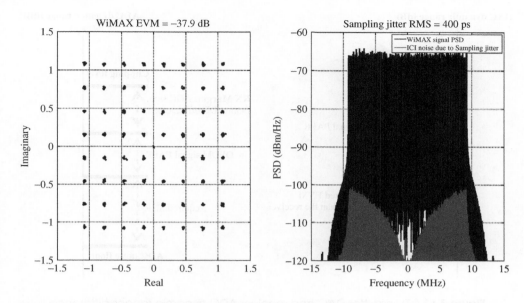

Figure 3.61   WiMAX 64-QAM constellation and ICI noise with sampling jitter of 400 ps

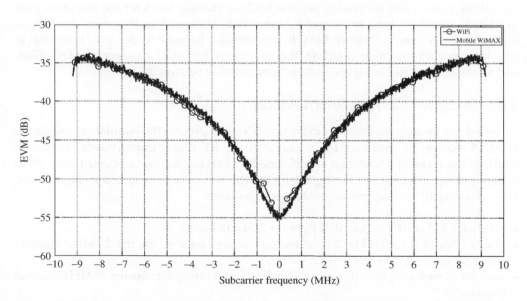

Figure 3.62   WiFi and WiMAX RX EVM per subcarrier for sampling jitter RMS of 400 ps

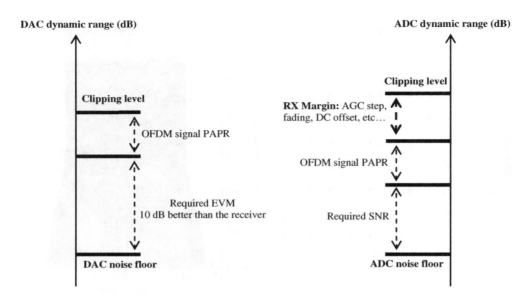

Figure 3.63   Transmitter DAC and receiver ADC dynamic ranges difference

simulation results and analysis of Section 3.5.2 are entirely valid for the reception path as was the case for phase noise and IQ mismatch. However, in reception the ADC is more difficult to specify than the DAC in transmission because its dynamic range has to take into account the channel fading, the interferers, the AGC accuracy, the DC offset, and so on, as depicted in Figure 3.63 and discussed in Section 2.8.

3.6.9 Receiver Complete Simulation

As was done for the transmission in Section 3.5.6, we present the simulation results for WiMAX receiver performance by including all the RF analog impairments previously studied. The receiver simulation is performed by varying the input power from −90 to 0 dBm and in each case estimating the EVM. The results are presented in Figure 3.64. The following RF analog impairments were assumed:

- residual CFO = 100 Hz (i.e., 0.04 ppm for 2.5 GHz band);
- SFO = 10 ppm (i.e., 224 Hz, 2% of the subcarrier spacing, for the 20 MHz channel using a sampling frequency of 22.4 MHz);
- receiver noise figure of 5 dB, giving −96 dBm input referred noise for 20 MHz channel bandwidth;
- LNA nonlinearity: IIP2 = 30 dBm, IIP3 = 10 dBm;
- IQ mismatch of 0.1 dB, 1°;
- phase noise: PLL bandwidth of 100 kHz, in-band noise of −95 dBc/Hz, giving an integrated phase noise of −43.1 dBc;

- sampling jitter of 400 ps RMS;
- ADC: 10 bits, clipping ratio of 12 dB.

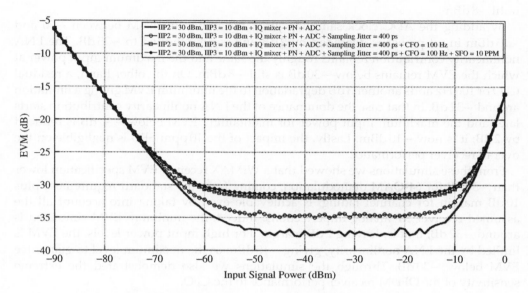

Figure 3.64 WiMAX receiver complete simulations

With respect to the performance target, the mobile WiMAX standard specifies for downlink reception an SNR higher than 20 dB, that is, EVM lower than −20 dB, for BER = 10^{-6} at the maximum data rate using a 64-QAM constellation (Table 3.3). However, this requirement is derived from simulations in AWGN; consequently, a margin has to be taken to account for real conditions including the propagation channel. Let us assume a margin of 10 dB specifying an EVM lower than −30 dB for the RF analog reception path.

Different simulation results are plotted in Figure 3.64 in order to incrementally quantify the impact of the RF analog front-end impairments on the receiver EVM.

In the first simulation we only took into account the receiver noise figure, the LNA IIP2/3, the IQ mismatch, the phase noise, and finally the ADC resolution and clipping. The EVM starts at −6 dB for −90 dBm input power due to the 5 dB noise figure, which gives a thermal noise floor of −96 dBm within the 20 MHz channel. The EVM decreases with increasing input power and starts to saturate around −37 dB above −50 dBm due to the receiver IQ mismatch and the phase noise. From the DAC/ADC simulation results plotted in Figure 3.10 we see that the 10-bit ADC contribution is negligible with an EVM smaller than −50 dB. It is interesting to note that this receiver EVM limitation is similar to the one obtained for the transmitter (simulation results depicted in Figure 3.30) because the IQ mismatch and the phase noise values are identical for both. This minimum EVM is guaranteed until −15 dBm, which is the input power above which the LNA nonlinearity starts to degrade the reception performance. Above −15 dBm the EVM is severely degraded by the LNA nonlinearity, which becomes the principal contributor, especially the IIP3, and an EVM lower than −30 dB is maintained until −8 dBm.

By adding the ADC clock sampling jitter, the minimum EVM between −50 and −15 dBm input power is degraded by 2 dB and is now limited to −35 dB. The LNA nonlinearity contribution remains roughly the same and the maximum input power at which the EVM remains below −30 dB is still −8 dBm. On the other hand, a residual CFO of 100 Hz adds another 3 dB degradation to the minimum EVM giving a limitation around −32 dB. In that case the dominance of the LNA nonlinearity contribution starts later and the maximum input power for maintaining EVM below −30 dB is reduced by 2 dB; it is now −10 dBm. Lastly, the impact of the 10 ppm SFO is negligible on the overall receiver performance.

From these simulations we showed that a WiMAX receiver EVM specification lower than −30 dB for 64-QAM demodulation, coming from the standard requirement plus 10 dB margin for channel fading, is achievable even by taking into account all the assumed receiver RF analog imperfections. Indeed, the minimum EVM simulated is around −32 dB; that is, leaving 2 dB margin. For high input power levels, the EVM is limited by the LNA nonlinearity, giving −10 dBm as the maximum level to guarantee EVM below −30 dB. Through the simulations we also demonstrated the extreme sensitivity of the OFDM receiver performance to the CFO.

3.7 Adaptive Modulation Illustration

In this brief section we aim to illustrate the impact of the receiver EVM on the different constellations used in Mobile WiMAX and the advantage of using adaptive modulation. The principle of adaptive modulation is to create a loop between the transmitter and the receiver in order to optimize the communication to the channel properties, that is, link quality. The goal is to allow the receiver to always demodulate the data at a specified BER by adapting the transmission modulation order, that is, the data-rate, to the received EVM.

For example, we show in Figure 3.65 the 64-QAM and 16-QAM constellations demodulated with an EVM around −15 dB. We can see that the 64-QAM constellation is difficult to distinguish and we can predict that the receiver will be unable to assure a BER of 10^{-6} with this modulation order because the EVM is below the minimum requirement defined in Table 3.3. On the other hand, by lowering the transmitter modulation order to 16-QAM, then for the same EVM the individual constellation points can be clearly observed allowing the receiver to achieve a BER of 10^{-6}, albeit at a lower data-rate.

Figure 3.66 shows the same principle when switching from 16-QAM to 4-QAM at the point where the EVM is such that the receiver is unable to correctly demodulate 16-QAM.

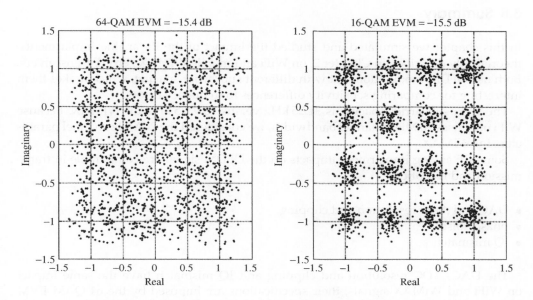

Figure 3.65 Adaptation of the modulation from 64-QAM to 16-QAM

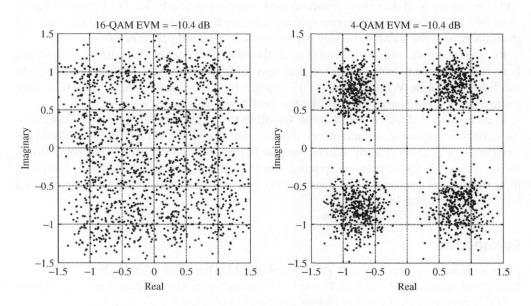

Figure 3.66 Adaptation of the modulation from 16-QAM to 4-QAM

3.8 Summary

In this chapter we simulated and studied the impact of the RF analog impairments, theoretically described in Chapter 2, on WiFi and mobile WiMAX zero-IF transceivers. Both use OFDM modulation, but with different subcarrier spacing, which makes them interesting for comparing sensitivity differences to frequency errors.

The WiFi subcarrier spacing is 312.5 kHz versus 10.9375 kHz for WiMAX. Because WiFi only uses 20 MHz channel bandwidth, we considered the 20 MHz channel parameters for WiMAX.

Some RF analog impairment impacts on the OFDM signal are comparable in transmission and reception:

- DAC and ADC resolution and clipping;
- phase noise;
- IQ mismatch.

The DAC/ADC resolution and clipping and IQ mismatch have the same impact on WiFi and WiMAX signals; their specifications are imposed by the 64-QAM EVM requirement. Regarding the EVM introduced by phase noise, because the WiFi subcarrier spacing is larger than typical RF PLL bandwidths used in modern transceivers, the CPE can be tracked and then compensated, improving the EVM. On the other hand, with WiMAX having a subcarrier spacing smaller than classical PLL bandwidths, ICI dominates and the EVM is directly given by the integrated phase noise.

In transmission, the PA distortion is the major bottleneck requiring a tradeoff between maximum output power and low EVM dictated by high-order 64-QAM modulation. Since WiMAX and WiFi signals have the same PAPR statistics, they impose the same linearity constraints for the PA in terms of back-off.

In reception, WiMAX is much more sensitive to carrier and SFOs than WiFi because its subcarrier spacing is narrower. Indeed, for the same carrier or SFO, WiMAX performance is around 30 dB worse because ICI noise power is inversely proportional to the square of the subcarrier spacing. On the other hand, the ADC clock sampling jitter introduces the same EVM degradation for WiFi and WiMAX because the channel bandwidths are similar.

References

Chen, K., Roberto, J., and Marco, B. (2008) *Mobile WiMAX*, John Wiley & Sons, Inc.

IEEE (1999) Standard 802.11a-1999. *Wireless LAN Medium Access Control (MAC) and Physical Layer (PHY) Specifications, High-Speed Physical Layer in the 5 GHz Band.*

IEEE (2003) Standard 802.11g-2003. *Wireless LAN Medium Access Control (MAC) and Physical Layer (PHY) Specifications. Further Higher-Speed Physical Layer Extension in the 2.4 GHz Band.*

IEEE (2005) Standard 802.16-2005. *Part 16: Air Interface for Fixed and Mobile Broadband Wireless Access Systems.*

Pollet, T., Spruyt, P., and Moeneclaey, M. (1994) The BER performance of OFDM systems using non-synchronized sampling. Proceedings Global Telecommunications Conference, 1994. GLOBECOM '94. *Communications: The Global Bridge*, vol. 1, 253–257.

Pollet, T., Van Bladel, M., and Moeneclaey, M. (1995) BER sensitivity of OFDM systems to carrier frequency offset and Wiener phase noise. *IEEE Transactions on Communications*, **43**, 191–193.

Rapp, C. (1991) Effects of HPA-nonlinearity on a 4-DPSK/OFDM-signal for a digital sound broadcasting system. Proceedings of the 2nd European Conference on Satellite Communications, pp. 176–184.

4

Digital Compensation of RF Analog Impairments

4.1 Introduction

With the ever-escalating demand for high-data rates for end-users, in addition to the use of larger frequency bandwidths, new wireless communication systems also specify higher order modulation, such as 64-QAM in mobile WiMAX and 3GPP-LTE, which necessitate high SNR in order to be properly demodulated. Consequently, these new high data-rate systems are particularly sensitive to the RF analog front-end impairments that are difficult to avoid in low-cost low-power transceivers integrated in deep sub-micrometer CMOS technology. Instead of trying to improve the RF analog blocks, which is neither cost nor power efficient, today the new trend in communication systems design is to accept increased RF analog front-end impairments, called dirty RF (Fettweis *et al.*, 2007), and to compensate the resulting errors in digital with advanced signal processing (Horlin and Bourdoux, 2008; Schenk, 2008).

In Chapters 2 and 3 we have described and simulated the major inevitable RF analog impairments we encounter in any practical communication system. These impairments are phase noise, mixer I and Q mismatch, amplifier nonlinearities, DAC and ADC clipping, sampling jitter, and carrier and SFOs). The aim was to analyze their impact on system performance with a special emphasis on OFDM modulation, which is widely used in modern communication standards (IEEE 802.11 a/g/n, 3GPP-LTE, Mobile WiMAX). We have seen that these impairments can severely degrade the link performance by generating in-band noise and distortion, due to the loss of subcarrier orthogonality, which limits the EVM (equivalently the SNR) and thus the maximum achievable data-rate. Because in OFDM transceivers the designer has access to both time and frequency domains, different estimation and compensation techniques have been proposed. In the time domain the schemes are typically based on auto- or

RF Analog Impairments Modeling for Communication Systems Simulation: Application to OFDM-based Transceivers, First Edition. Lydi Smaini.
© 2012 John Wiley & Sons, Ltd. Published 2012 by John Wiley & Sons, Ltd.

inter-correlation processing using training symbols sent before the data-symbols, or the repetitive cyclic prefix. In the frequency domain, subcarrier-pilots with known phase and amplitude are used to estimate and track the impairments.

We will study in this chapter how some of these RF analog imperfections can be estimated and compensated in DBB. Before describing the correction schemes it is important to understand the impact on system performance; therefore, we will review their effects based on the theoretical and simulation results presented in the two previous chapters.

We will first concentrate on carrier and SFOs which are important RF impairments, especially for OFDM-based transceivers, which are very sensitive to these two synchronization errors caused by oscillator frequency mismatch between transmission and reception. Afterwards, we will discuss I and Q mismatch in zero-IF architecture and a method for estimating the complex conjugated image in the frequency domain, in order to derive compensation coefficients for reducing the quadrature mixer imbalance effect. Finally, we will present a signal processing technique used in transmission for compensating the PA nonlinearity by distorting the baseband signal before the PA.

4.2 CFO Estimation and Correction

4.2.1 CFO Estimation Principle

We have seen in Section 2.5 that the main issue with CFO is that it introduces two distinct components which degrade the overall system performance. For each subcarrier the two deleterious effects are

- Phase rotation and amplitude reduction common to all the subcarriers.
- ICI caused by the other subcarriers.

In addition, from one OFDM symbol to the next CFO also introduces another deterministic phase increment to the subcarriers depending on the OFDM symbol duration. Taking the first symbol as the reference, $i = 1$, the phase rotation introduced by CFO for symbol i is

$$\Delta\phi(i) = \pi \frac{\Delta_{\text{CFO}}}{\Delta f} + 2\pi(i - 1)\Delta_{\text{CFO}}N_{\text{s}}T_{\text{s}} \tag{4.1}$$

where i is symbol index, Δ_{CFO} is the CFO in hertz, Δf is the subcarrier spacing in hertz, $N_{\text{s}} = N_{\text{g}} + N$ is the number of points of a single OFDM symbol including the guard interval, and T_{s} is the sampling frequency.

Figure 4.1 is a plot of Equation 4.1 showing that the phase shift introduced by CFO is a linear function of the symbol index. The slope of the phase shift is given by

$$\frac{\mathrm{d}\Delta\phi(i)}{\mathrm{d}i} = 2\pi\,\Delta_{\text{CFO}}N_{\text{s}}T_{\text{s}} \tag{4.2}$$

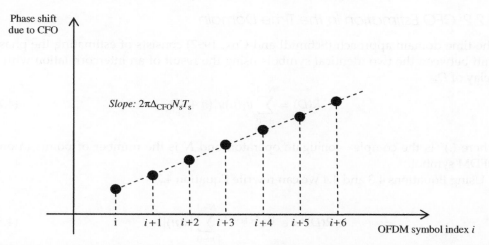

Figure 4.1 Illustration of the inter-symbol phase shift introduced by carrier frequency offset

From Equations 4.1 and 4.2 we can see that the only unknown in the common phase rotation is the term Δ_{CFO} related to the CFO which we want to estimate and correct.

The principle of the CFO estimation method is based on estimating the change in phase between OFDM symbols, that is, the slope, in order to extract Δ_{CFO}.

Let us suppose two identical OFDM symbols $x(n)$ received with a time delay D affected by the same CFO Δ_{CFO}:

$$
\begin{aligned}
y_1(n) &= x(n)\, \mathrm{e}^{\mathrm{j}2\pi\,\Delta_{\text{CFO}} n T_{\mathrm s}} \\
y_2(n) &= x(n)\, \mathrm{e}^{\mathrm{j}2\pi\,\Delta_{\text{CFO}} (n+D) T_{\mathrm s}}
\end{aligned}
\tag{4.3}
$$

We can rewrite Equation 4.3 as

$$
\begin{aligned}
y_1(n) &= x(n)\, \mathrm{e}^{\mathrm{j}2\pi\,\Delta_{\text{CFO}} n T_{\mathrm s}} \\
y_2(n) &= y_1(n)\, \mathrm{e}^{\mathrm{j}2\pi\,\Delta_{\text{CFO}} D T_{\mathrm s}}
\end{aligned}
\tag{4.4}
$$

showing that the only difference between the two symbols is simply a deterministic phase shift:

$$
\Delta\phi = 2\pi\,\Delta_{\text{CFO}} D T_{\mathrm s}
\tag{4.5}
$$

The CFO is extracted from Equation 4.5:

$$
\Delta_{\text{CFO}} = \frac{\Delta\phi}{2\pi D T_{\mathrm s}}
\tag{4.6}
$$

4.2.2 CFO Estimation in the Time Domain

The time domain approach (Schmidl and Cox, 1997) consists of estimating the phase shift between the two identical symbols using the result of an intercorrelation with a delay of D:

$$R(D) = \sum_{n=0}^{N-1} y_1(n) y_1^*(n+D) \tag{4.7}$$

where $(.)^*$ is the complex conjugate operator, and N is the number of points in one OFDM symbol.

Using Equations 4.3 and 4.4 we can rewrite Equation 4.7:

$$R(D) = e^{-j2\pi \Delta_{CFO} D T_s} \sum_{n=0}^{N-1} |x(n)|^2 \tag{4.8}$$

which is a complex variable whose angle is proportional to the CFO:

$$\angle R(D) = -2\pi \Delta_{CFO} D T_s \tag{4.9}$$

The CFO estimator is then given by:

$$\Delta_{CFO} = -\frac{\angle R(D)}{2\pi D T_s} \tag{4.10}$$

Because the unambiguous range for the angle of Equation 4.9 is only defined between $+\pi$ and $-\pi$, the maximum absolute carrier frequency error that can be estimated is

$$|\Delta_{CFO}| \leq \frac{\pi}{2\pi D T_s} = \frac{1}{2 D T_s} \tag{4.11}$$

Because the sampling frequency, that is, T_s, is generally fixed, the maximum frequency offset defined by Equation 4.11 is fixed by the delay D which defines the training symbol length. In practice the CFO estimation is decomposed into two consecutive operations: coarse and fine estimation. In coarse estimation, short training symbols are used for covering a large CFO range; afterwards, long training symbols are sent in order to refine the estimate.

Figure 4.2 depicts the time-domain CFO estimator defined by Equation 4.7 using first short training symbols $x_s(n)$ followed by long training symbols $x_L(n)$. The delay D must be adapted to the training symbol length.

We have performed simulations in order to test the performance of this time-domain CFO estimator as a function of the SNR fixed by an AWGN channel. We assumed a CFO of 10 kHz, a training sequence composed of two identical OFDM symbols with $N = 128$ subcarriers, a sampling frequency of 20 MHz, and SNR varying from 0 to 50 dB. The value of the delay D is $N = 128$; as a result, we can derive from Equation 4.11 the

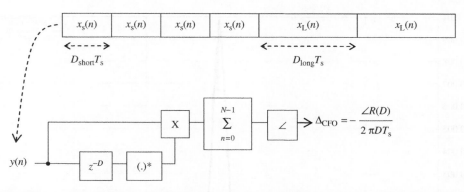

Figure 4.2 Carrier frequency estimator in the time domain with short and long training symbols

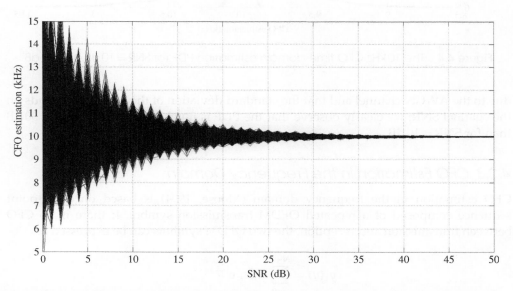

Figure 4.3 The 10 kHz CFO time-domain estimation as a function of SNR (10 000 realizations)

maximum carrier frequency error that can be estimated, which is 78.125 kHz and is larger than the 10 kHz assumption.

Figure 4.3 shows 10 000 realizations of the CFO estimation as a function of the SNR. We can see that for low SNR, lower than 10 dB, the error on the estimation is non-negligible compared with the CFO, exceeding 10%. As the SNR increases, the estimation becomes more accurate, as expected.

The probability density functions (PDFs) of the CFO estimate at SNR = 10 dB and 30 dB are depicted in Figure 4.4. We can notice that both PDFs have a Gaussian shape

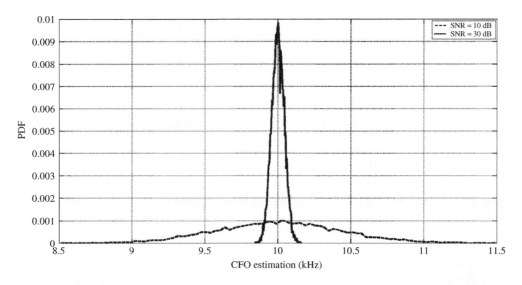

Figure 4.4 The 10 kHz CFO time-domain estimation PDF for SNR = 10 dB and 30 dB

due to the AWGN channel and that the standard deviation of the estimate depends on the channel SNR; we clearly observe that the PDF is much narrower for SNR = 30 dB than for SNR = 10 dB.

4.2.3 CFO Estimation in the Frequency Domain

CFO estimation in the frequency domain (Moose, 1994) is based on a 2N-point sequence composed of a repeated OFDM transmission symbol. If there is no CFO between transmission and reception, the two OFDM symbols can be expressed as

$$
y_1(n) = \sum_{k=0}^{N-1} H_{1k} s_k \, e^{j\frac{2\pi}{N} kn}
$$

$$
y_2(n) = \sum_{k=0}^{N-1} H_{2k} s_k \, e^{j\frac{2\pi}{N} k(n+N)}
$$

(4.12)

where $n = 0 \ldots N-1$, s_k are the data symbols, k is the subcarrier index, N is the total number of subcarriers, and H_{1k} and H_{2k} are the channel frequency responses for symbols 1 and 2, respectively.

In the case of a slowly varying channel, which is a valid assumption if we assume two consecutive OFDM symbols, we have $H_{1k} \approx H_{2k}$.

In the presence of CFO, Equation 4.12 becomes

$$y_1(n) = \sum_{k=0}^{N-1} H_{1k} s_k \, e^{j\frac{2\pi}{N}\left(k+\frac{\Delta_{\text{CFO}}}{\Delta f}\right)n}$$

$$y_2(n) = \sum_{k=0}^{N-1} H_{1k} s_k \, e^{j\frac{2\pi}{N}\left(k+\frac{\Delta_{\text{CFO}}}{\Delta f}\right)(n+N)} = y_1(n) \, e^{j2\pi\frac{\Delta_{\text{CFO}}}{\Delta f}}$$

(4.13)

in which Δ_{CFO} is CFO and Δf is the subcarrier spacing, both in hertz.

In reception, after applying the DFT on the symbols we obtain

$$Y_1(m) = \frac{1}{N} \sum_{n=0}^{N-1} y_1(n) \, e^{-j\frac{2\pi}{N}mn}$$

$$Y_2(m) = \frac{1}{N} \sum_{n=0}^{N-1} y_2(n) \, e^{-j\frac{2\pi}{N}mn} = \frac{1}{N} \sum_{n=0}^{N-1} y_1(n) \, e^{-j\frac{2\pi}{N}mn} \, e^{j2\pi\frac{\Delta_{\text{CFO}}}{\Delta f}} \, e^{j2\pi k}$$

(4.14)

from which we can derive this interesting result for the considered subcarrier $m = k$:

$$Y_2(k) = Y_1(k) \, e^{j2\pi\frac{\Delta_{\text{CFO}}}{\Delta f}}$$

(4.15)

Equation 4.15 shows that the two DFT results are equal in amplitude, but all the subcarriers are affected by a common phase shift directly proportional to the normalized CFO to the subcarrier spacing.

The estimation of the frequency offset is then performed by measuring the phase difference between the two DFT results using a maximum likelihood estimator (Moose, 1994):

$$\theta = \tan^{-1}\left\{ \frac{\displaystyle\sum_{k=0}^{N-1} \text{Im}\left[Y_2(k)Y_1^*(k)\right]}{\displaystyle\sum_{k=0}^{N-1} \text{Re}\left[Y_2(k)Y_1^*(k)\right]} \right\} = 2\pi\frac{\Delta_{\text{CFO}}}{\Delta f}$$

(4.16)

From this we can extract an estimate of the relative CFO normalized to the subcarrier spacing:

$$\frac{\Delta_{\text{CFO}}}{\Delta f} = \frac{\theta}{2\pi}$$

(4.17)

and finally the real CFO in hertz:

$$\Delta_{\text{CFO}} = \frac{\theta}{2\pi}\Delta f$$

(4.18)

Because the estimate of the angle θ is limited between $+\pi$ and $-\pi$, the maximum absolute carrier frequency error that can be measured is half of the subcarrier spacing:

$$-\frac{\Delta f}{2} \leq \Delta_{\text{CFO}} < \frac{\Delta f}{2} \tag{4.19}$$

For CFOs greater than one half of the subcarrier spacing, an initial estimation has to be done before applying this estimator. This is the major limitation of the frequency-domain method compared with the time-domain approach, in which the maximum frequency estimation can be adjusted with the training symbol length.

Compared with the time-domain estimator, the main difference of the frequency-domain estimation of the CFO is the processing of the DFT for both repeated symbols leading to additional computations without significant performance gain.

In order to compare the performance of the CFO estimation in the frequency domain versus the time-domain estimator described in the previous section, we used the same configuration to study its sensitivity to the SNR fixed by an AWGN channel. We again assumed a CFO of 10 kHz, a training sequence composed of two identical OFDM symbols with $N = 128$ subcarriers, a sampling frequency of 20 MHz, and SNR varying from 0 to 50 dB. We can see from Figures 4.5 and 4.6 that the results of the CFO estimation in the frequency domain are similar to those obtained in the time domain, which are represented in Figures 4.3 and 4.4. Actually, the results are identical because the sensitivity of the algorithm to the noise is a function of the noise bandwidth, or equivalently the OFDM symbol length.

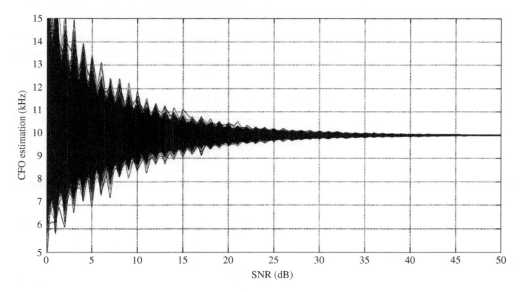

Figure 4.5 The 10 kHz CFO frequency-domain estimation as a function of SNR (10 000 realizations)

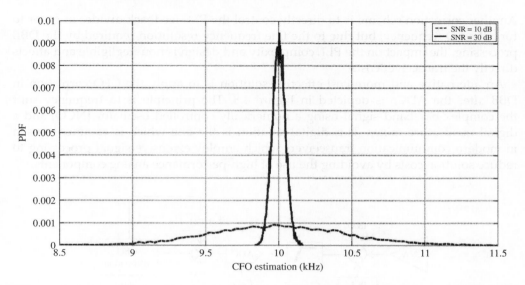

Figure 4.6 The 10 kHz CFO frequency-domain estimation PDF for SNR = 10 dB and 30 dB

4.2.4 CFO Correction

In the previous section we described two CFO estimation methods based on measurements of the phase difference between two identical OFDM symbols, either in the time domain or in the frequency domain. Regarding the correction, if only the subcarrier phase shift is compensated, the ICI will remain and the reception performance will not be improved. An efficient correction is to compensate the CFO before applying the DFT by frequency shifting the received signal.

The carrier frequency correction can be applied to the RF signal by adjusting the reference oscillator, voltage controlled crystal oscillator (VCXO), of the PLL LO used for quadrature frequency down-conversion, as illustrated in Figure 4.7. Although this correction method is straightforward, it requires a controllable reference oscillator which increases the overall transceiver cost and also requires dedicated control logic.

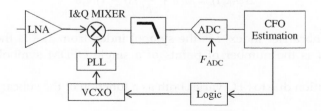

Figure 4.7 CFO correction in RF by shifting the PLL reference frequency

Another correction scheme is to directly control the PLL feedback divider in order to tune the LO frequency; but due to the fine frequency resolution required by the DBB processing, the impact on the PLL complexity and design is non-negligible and affects directly the transceiver cost.

An alternative and more cost-effective solution is to apply the CFO correction in DBB after the ADC, as depicted in Figure 4.8. The principle is to frequency shift the complex baseband signal using a numerically controlled oscillator (NCO) and a digital quadrature mixer. This digital solution is the one which is commonly used in modern communication transceivers which employ extensive signal processing to reduce solution costs by avoiding the use of high-performance analog components.

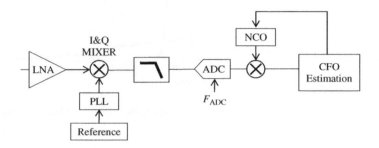

Figure 4.8 CFO correction applied on the digital complex baseband signal

4.3 SFO Estimation and Correction

4.3.1 SFO Estimation Principle

We have seen in Section 2.6 that SFO, as for CFO, has a twofold effect with a phase rotation and an amplitude reduction for all the subcarriers, and ICI. However, there is a fundamental difference, because the phase rotation and amplitude reduction are not common for all the subcarriers but instead depend on the subcarrier index. The phase rotation introduced by SFO is given by

$$\Delta\phi(k, i) = k\pi\delta + \frac{2\pi}{N}k(i - 1)N_s\delta \tag{4.20}$$

where k is the subcarrier index, i is the symbol index from 1, δ is the relative SFO, and $N_s = N_g + N$ is the number of points of a single OFDM symbol including the guard interval.

The phase rotation due to SFO varies both as a function of the subcarrier index:

$$\frac{\mathrm{d}\Delta\phi(k, i)}{\mathrm{d}k} = \pi\delta + \frac{2\pi}{N}(i - 1)N_s\delta \tag{4.21}$$

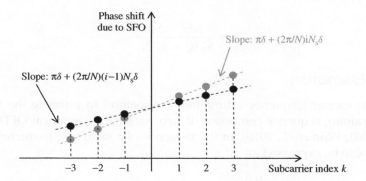

Figure 4.9   Phase shift due to SFO as a function of the subcarrier index

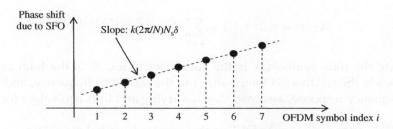

Figure 4.10   Single subcarrier phase shift due to SFO as a function of the OFDM symbol index

and as a function of the OFDM symbol index by

$$\frac{\mathrm{d}\Delta\phi(k,i)}{\mathrm{d}i} = \frac{2\pi}{N}kN_\mathrm{s}\delta \tag{4.22}$$

Equations 4.21 and 4.22 are plotted in Figures 4.9 and 4.10, respectively. Figure 4.9 illustrates that the phase rotation due to SFO increases linearly versus the subcarrier index, with a slope dependent on the symbol index, as demonstrated in Equations 4.20 and 4.21. Consequently, it is difficult to estimate the SFO without knowing the phase shift and symbol position relative to those of a perfect OFDM symbol timing reference. On the other hand, Figure 4.10 represents the phase rotation due to SFO for one subcarrier as a function of the OFDM symbol index. In this case we can see that the phase rotation slope depends on the subcarrier index. The major SFO estimation algorithms use these two properties for extracting the SFO using the phase differences between two subcarriers across two OFDM symbols:

$$\Delta\phi_1 = \Delta\phi(k_1, i+1) - \Delta\phi(k_1, i) = \frac{2\pi}{N}k_1 N_\mathrm{s}\delta$$

$$\Delta\phi_2 = \Delta\phi(k_2, i+1) - \Delta\phi(k_2, i) = \frac{2\pi}{N}k_2 N_\mathrm{s}\delta \tag{4.23}$$

giving

$$\delta = \frac{\Delta\phi_2 - \Delta\phi_1}{\frac{2\pi}{N}(k_2 - k_1)N_s} \tag{4.24}$$

4.3.2 SFO Estimation

Similar to the carrier frequency estimation, the method to estimate the SFO is also based on a training sequence composed of two consecutive identical OFDM symbols (Sliskovic, 2001; Won *et al.*, 2010). In the presence of a sampling frequency error, the two symbols can be expressed as

$$y_1(n) = \sum_{k=0}^{N-1} H_k s_k \, e^{j\frac{2\pi}{N}kn(1+\delta)}$$

$$y_2(n) = y_1(n+N) = \sum_{k=0}^{N-1} H_k s_k \, e^{j\frac{2\pi}{N}k(n+N)(1+\delta)} \tag{4.25}$$

where s_k are the data symbols, k is the subcarrier index, N is the total number of subcarriers, δ is the relative SFO normalized to the sampling frequency, and H_k is the channel frequency response, assumed slowly varying and thus equivalent for symbols 1 and 2.

In reception, after applying the DFT on this repeated symbol we obtain

$$Y_1(m) = \frac{1}{N}\sum_{n=0}^{N-1} y_1(n)\, e^{-j\frac{2\pi}{N}mn} = \frac{1}{N}\sum_{k=0}^{N-1} H_k s_k \sum_{n=0}^{N-1} e^{j\frac{2\pi}{N}(k-m)n\delta}\, e^{j\frac{2\pi}{N}kn\delta}$$

$$Y_2(m) = \frac{1}{N}\sum_{n=0}^{N-1} y_2(n)\, e^{-j\frac{2\pi}{N}mn} = \frac{1}{N}\sum_{k=0}^{N-1} H_k s_k \, e^{j2\pi k\delta}e^{j2\pi k}\sum_{n=0}^{N-1} e^{j\frac{2\pi}{N}(k-m)n\delta}\, e^{j\frac{2\pi}{N}kn\delta} \tag{4.26}$$

In Section 2.6.2 we demonstrated that the SFO introduces a phase rotation and an amplitude attenuation of the $m = k$ subcarrier considered and ICI noise for $m \neq k$, which limits the performance. The principle of the SFO estimation method is to measure and track the phase rotation of the considered subcarrier $m = k$.

As a result, with $m = k$ Equation 4.26 becomes

$$Y_1(m = k) = H_k s_k \sum_{n=0}^{N-1} e^{j\frac{2\pi}{N}kn\delta}$$

$$Y_2(m = k) = H_k s_k \, e^{j2\pi k\delta}\sum_{n=0}^{N-1} e^{j\frac{2\pi}{N}kn\delta} \tag{4.27}$$

From Equation 4.27 we can notice that the difference between the spectra of these two consecutive identical OFDM symbols is a phase shift of the data symbols s_k depending on the relative SFO and linearly increasing with the subcarrier index:

$$Y_2(k) = Y_1(k)\, e^{j2\pi k\delta} \tag{4.28}$$

The SFO estimate is calculated by using the difference in this symbol phase shift between two subcarriers spaced by Δk:

$$\Delta\theta = \tan^{-1}\left\{ \frac{\mathrm{Im}\left[Y_1(k)Y_2^*(k)\right]}{\mathrm{Re}\left[Y_1(k)Y_2^*(k)\right]} \right\} - \tan^{-1}\left\{ \frac{\mathrm{Im}\left[Y_1(k+\Delta k)Y_2^*(k+\Delta k)\right]}{\mathrm{Re}\left[Y_1(k+\Delta k)Y_2^*(k+\Delta k)\right]} \right\} \tag{4.29}$$

giving

$$\Delta\theta = -2\pi k\delta + 2\pi(k+\Delta k)\delta = 2\pi\,\Delta k\delta \tag{4.30}$$

From this we obtain an estimate of the relative SFO:

$$\delta = \frac{\Delta\theta}{2\pi\,\Delta k} \tag{4.31}$$

and finally the real SFO in hertz:

$$\mathrm{SFO} = \frac{\Delta\theta}{2\pi\,\Delta k}F_s \tag{4.32}$$

where F_s is the sampling frequency in hertz.

In order to have a better estimate, especially in the presence of noise which degrades any practical communication system performance, a weighted averaging can be performed:

$$\frac{1}{L}\sum_{k=1}^{L} w_k\delta_k \tag{4.33}$$

where w_k are the weighting factors, L is the number of estimations, and δ_k is the SFO estimate from subcarriers k and $k+\Delta k$.

By assuming comparable SNR for close subcarriers k and $k+\Delta k$, Sliskovic (2001) proposes to use the normalized SNR per subcarrier as weighting factor:

$$w_k = \frac{\mathrm{SNR}_k}{\frac{1}{L}\sum_{k=1}^{L}\mathrm{SNR}_k} \tag{4.34}$$

Figures 4.11 and 4.12 present sampling frequency estimation results as a function of the SNR as it varies from 10 to 50 dB in AWGN. We supposed an SFO of 100 ppm, a training sequence composed of two identical OFDM symbols with $N = 128$ subcarriers, and an ideal sampling frequency of 20 MHz. In Figure 4.11 we can see that the estimate accuracy improves and tends toward 100 ppm as the SNR increases in the same way,

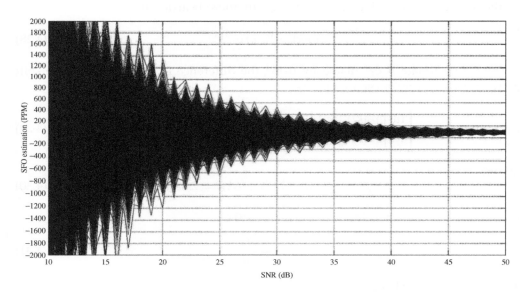

Figure 4.11   The 100 ppm SFO estimation as a function of SNR (10 000 realizations)

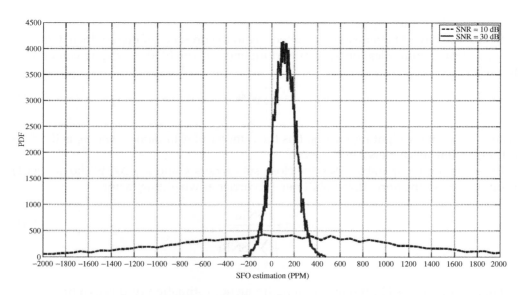

Figure 4.12   The 100 ppm SFO estimation PDF for SNR = 10 dB and 30 dB

as for CFO estimation. This is confirmed by the two PDFs plotted in Figure 4.12 for SNR = 10 dB and SNR = 30 dB, both of which are centered on 100 ppm. At the same time we observe that the estimation performs poorly even at SNR = 10 dB, as confirmed by the large standard deviation of the PDF, whereas it was much better for CFO estimation. This highlights that sampling frequency estimation is much more sensitive to noise than carrier frequency estimation, as we will see later with joint carrier and SFO estimation.

4.3.3 SFO Correction

The receiver SFO can be corrected and tracked using synchronous or non-synchronous techniques. In synchronized sampling systems the principle consists of directly tuning the ADC clock frequency in order to align the receiver sampling clock with that of the transmitter, as depicted in Figure 4.13. Alternatively, in non-synchronized sampling systems the ADC clock remains constant and the SFO is compensated in digital with interpolation filtering before the FFT processing (Figure 4.14). Similar to CFO correction, in modern transceivers in which digital signal processing is emphasized, the non-synchronous scheme is preferred because it is simpler to implement and more cost effective with the avoidance of a controllable oscillator and expensive external analog components.

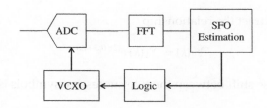

Figure 4.13   ADC sampling frequency correction: synchronized sampling

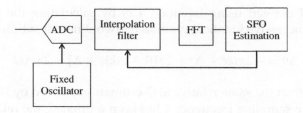

Figure 4.14   SFO correction in digital after the ADC: non-synchronized sampling

4.3.4 Joint SFO and CFO Estimation

The method previously described to estimate the SFO from two repeated OFDM symbols can also be used to extract the CFO (Sliskovic, 2001; Lei *et al.*, 2003). In the presence of carrier and SFOs, Equation 4.25 can be rewritten

$$y_1(n) = \sum_{k=0}^{N-1} H_k s_k \, e^{j\frac{2\pi}{N}[k(1+\delta)+\Delta]n}$$

$$y_2(n) = y_1(n+N) = \sum_{k=0}^{N-1} H_k s_k \, e^{j\frac{2\pi}{N}[k(1+\delta)+\Delta](n+N)}$$

(4.35)

where s_k are the data symbols, N is the total number of subcarriers, δ is the relative SFO normalized to the sampling frequency, Δ is the relative CFO normalized to the carrier frequency, and H_k is the channel frequency response, assumed slowly varying and thus equivalent for symbols 1 and 2.

After DFT processing in reception, for the $m = k$ subcarrier considered we obtain

$$Y_1(m = k) = \frac{1}{N}\sum_{n=0}^{N-1} y_1(n)\, e^{-j\frac{2\pi}{N}mn} = H_k s_k \sum_{n=0}^{N-1} e^{j\frac{2\pi}{N}(k\delta+\Delta)n}$$

$$Y_2(m = k) = \frac{1}{N}\sum_{n=0}^{N-1} y_2(n)\, e^{-j\frac{2\pi}{N}mn} = H_k s_k \, e^{j2\pi(k\delta+\Delta)} \sum_{n=0}^{N-1} e^{j\frac{2\pi}{N}(k\delta+\Delta)n}$$

(4.36)

from which we can extract the relationship

$$Y_2(k) = Y_1(k)\, e^{j2\pi(k\delta+\Delta)}$$

(4.37)

showing that the phase shift between the two successive symbols is

$$\tan^{-1}\left\{\frac{\mathrm{Im}\left[Y_1(k)Y_2^*(k)\right]}{\mathrm{Re}\left[Y_1(k)Y_2^*(k)\right]}\right\} = -2\pi(k\delta+\Delta)$$

(4.38)

The SFO is still derived using Equation 4.29 by calculating the phase difference between the two successive identical symbols on two different subcarriers:

$$\Delta\theta = -2\pi(k\delta+\Delta) + 2\pi[(k+\Delta k)\delta + \Delta] = 2\pi\,\Delta k\delta$$

(4.39)

from which we extract the same relative SFO estimate expressed by Equation 4.31.

Once the relative sampling frequency δ has been estimated, the relative CFO can be derived for each subcarrier from Equation 4.38:

$$\Delta_k = -\frac{1}{2\pi}\tan^{-1}\left\{\frac{\mathrm{Im}\left[Y_1(k)Y_2^*(k)\right]}{\mathrm{Re}\left[Y_1(k)Y_2^*(k)\right]}\right\} - k\delta$$

(4.40)

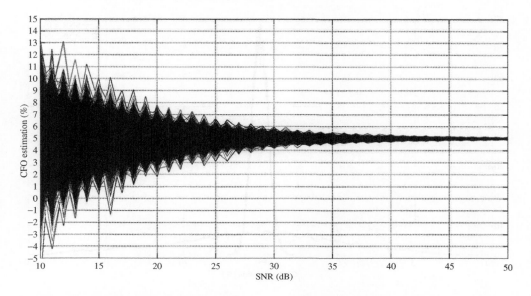

Figure 4.15 The 5% CFO estimation from SFO estimation as a function of SNR

As for SFO estimation, the accuracy can be improved by performing an averaging over all subcarriers using the same weighting factors defined in Equation 4.33:

$$\Delta = \frac{1}{L} \sum_{k=1}^{L} w_k \Delta_k \tag{4.41}$$

where w_k are the weighting factors, L is the number of estimations, and Δ_k is the CFO estimate from subcarriers k and $k + \Delta k$.

Figures 4.15 and 4.16 show simulation results for CFO estimated jointly with SFO as a function of SNR. The simulation configuration is similar to the previous case with an SFO of 100 ppm, but in addition the signal is affected by a CFO of 5% of the subcarrier spacing; that is, 5% of 20 MHz/128 = 7.8125 kHz. We can clearly see that the CFO is well estimated with the PDFs of Figure 4.16 centered on 5%, and as expected the standard deviation decreases as the SNR increases.

4.4 IQ Mismatch Estimation and Correction

4.4.1 Principle

We have seen in Section 2.7 that the quadrature mixer amplitude and phase mismatches can severely degrade zero-IF OFDM transceiver performance by generating a complex

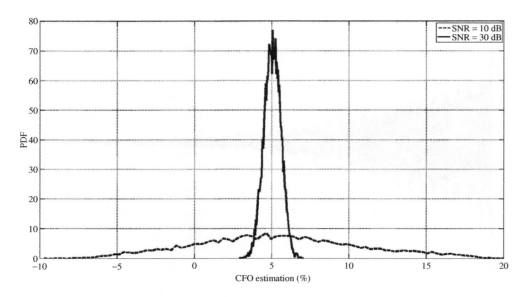

Figure 4.16 The 5% CFO estimation from SFO estimation PDF for SNR = 10 and 30 dB

conjugate image superimposed on the original signal. We demonstrated that the complex baseband signal including the IQ mismatch effect on the ideal signal $x(t)$ can be expressed as

$$y_{BB}(t) = \alpha x(t) + \beta x^*(t) \tag{4.42}$$

in which the ratio β/α is the ISR of the quadrature mixer, with

$$\alpha = \frac{1}{2}\left[\cos(\Delta\phi/2) - j\frac{\Delta G}{2}\sin(\Delta\phi/2)\right]$$
$$\beta = \frac{1}{2}\left[-\frac{\Delta G}{2}\cos(\Delta\phi/2) + j\sin(\Delta\phi/2)\right] \tag{4.43}$$

where ΔG and $\Delta\phi$ are the amplitude and phase imbalances, respectively, and $(.)^*$ denotes the complex conjugate.

Let us suppose an ideal OFDM symbol not affected by the channel propagation:

$$x(n) = \sum_{k=0}^{N-1} s_k \, e^{j\frac{2\pi}{N}kn} \tag{4.44}$$

where s_k are the data symbols, k is the subcarrier index, and N is the total number of subcarriers.

Applying the IQ mismatch from Equation 4.42, we obtain

$$y_{BB}(n) = \alpha x(n) + \beta x^*(n) = \alpha \sum_{k=0}^{N-1} s_k\, e^{j\frac{2\pi}{N}kn} + \beta \sum_{k=0}^{N-1} s_k^*\, e^{-j\frac{2\pi}{N}kn} \tag{4.45}$$

In reception, after the DFT processing we obtain the impaired signal in frequency domain (Tubbax *et al.*, 2003, 2004):

$$Y_{BB}(m = k) = \frac{1}{N} \sum_{n=0}^{N-1} y_{BB}(n)\, e^{-j\frac{2\pi}{N}mn} = \alpha X(k) + \beta X^*(-k) = \alpha s_k + \beta s_{-k}^* \tag{4.46}$$

showing that the ideal symbols s_k are corrupted by the complex conjugate of symbol s_{-k}, as illustrated in Figure 4.17. If the symbols s_k and s_{-k} are known, for example, via a training sequence or pilot-tones sent before the data, Tubbax *et al.* (2003) proposed a frequency-domain correction technique derived from Equation 4.46:

$$s_k = \frac{Y_{BB}(k) - \beta s_{-k}^*}{\alpha} \tag{4.47}$$

Equation 4.47 shows that if one can estimate the coefficients α and β, it is possible to compensate the effect of IQ mismatch because we make the assumption that $Y_{BB}(k)$ and s_{-k} are known.

In order to estimate the two unknowns, α and β, we require at least two subcarriers to produce two linear equations:

$$\begin{aligned}
Y_{BB}(k) &= \alpha s_k + \beta s_{-k}^* \\
Y_{BB}(l) &= \alpha s_l + \beta s_{-l}^*
\end{aligned} \tag{4.48}$$

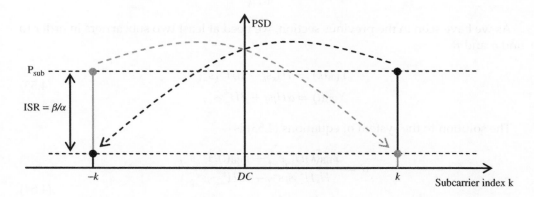

Figure 4.17 Effect of IQ mismatch in the frequency domain for OFDM transceivers

Equation 4.48 can be easily solved:

$$\alpha = \frac{Y_{BB}(k)s^*_{-l} - Y_{BB}(l)s^*_{-k}}{s_k s^*_{-l} - s_l s^*_{-k}}$$

$$\beta = \frac{Y_{BB}(l) - \alpha s_l}{s^*_{-l}} \tag{4.49}$$

Several pairs of subcarriers can be used in order to obtain more than one estimate of the coefficients α and β, and afterwards an average can be computed for use in the compensation Equation 4.47.

4.4.2 Effect of the Channel

If we take into account the channel, Equation 4.46 becomes

$$Y_{BB}(k) = \alpha H_k s_k + \beta H^*_{-k} s^*_{-k} \tag{4.50}$$

where H_k is the channel coefficient in the frequency domain for subcarrier index k.

After channel estimation and equalization using a zero-forcing equalizer (Section 1.3.3) we obtain an estimate of the symbol s_k biased by the conjugate of the symbol s_{-k} and the channel coefficients:

$$\frac{Y_{BB}(k)}{H_k} = \alpha s_k + \beta \frac{H^*_{-k} s^*_{-k}}{H_k} \tag{4.51}$$

From Equation 4.51, Tubbax *et al.* (2003) propose a compensation scheme similar to the one defined by Equation 4.47:

$$s_k = \frac{Y_{BB}(k) - \beta H^*_{-k} s^*_{-k}}{\alpha H_k} \tag{4.52}$$

As we have seen in the previous section, we need at least two subcarriers in order to find α and β:

$$Y_{BB}(k) = \alpha H_k s_k + \beta H^*_{-k} s^*_{-k}$$

$$Y_{BB}(l) = \alpha H_l s_l + \beta H^*_{-l} s^*_{-l} \tag{4.53}$$

The solution to the system of equations (4.53) is

$$\alpha = \frac{Y_{BB}(k)H^*_{-l}s^*_{-l} - Y_{BB}(l)H^*_{-k}s^*_{-k}}{H_k H^*_{-l}s_k s^*_{-l} - H_l H^*_{-k}s_l s^*_{-k}}$$

$$\beta = \frac{Y_{BB}(l) - \alpha H_l s_l}{H^*_{-l}s^*_{-l}} \tag{4.54}$$

which is fairly similar to Equation 4.49 for the coefficient estimation without the channel effect. The main difference is the scaling of symbols by the channel transfer function that is generally estimated with α and β since they cannot distinguished separately.

4.4.3 Simulation Results

In order to study the performance of the proposed IQ mismatch estimation and compensation scheme, we included the algorithm in the WiMAX receiver simulation chain described in Chapter 3. We computed the ISR coefficients α and β using the pilots which are part of the 2048 subcarriers OFDM symbol, and then these coefficients have been applied across all the subcarriers for compensating the receiver IQ mismatch.

We assumed receiver gain and phase imbalance of 1 dB (12.2%) and 3°, respectively, giving a theoretical ISR of -23.6 dB (Section 2.7).

Figure 4.18 shows the receiver EVM estimation with and without IQ mismatch compensation as a function of SNR as it varies from 0 to 50 dB fixed by an AWGN channel. In addition we plotted the ISR computed from the coefficients α and β estimated with two pilots. We can see that the ISR estimation is very noisy for low SNR because it is corrupted and limited by the AWGN channel, but it tends toward the theoretical value of -23.6 dB as the SNR increases. It is interesting to note that without IQ mismatch compensation the EVM is limited by the ISR as expected. On the other hand, with IQ compensation the EVM results are enhanced and we can observe that they are not limited by the ISR anymore and the EVM can reach -40 dB

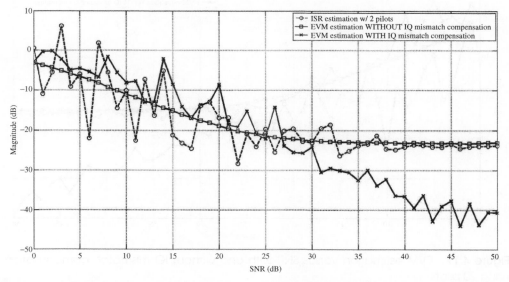

Figure 4.18 EVM estimation versus SNR with and without IQ mismatch compensation using two pilots

for SNR higher than 40 dB, which is around 16 dB below the ICI noise generated by the IQ mismatch.

In the simulation results depicted in Figure 4.19 the IQ mismatch compensation coefficients α and β are estimated using 20 pilots in order to obtain a better accuracy using an average value. The results are quite similar to the previous ones; however, we can observe that the EVM estimation with the compensation is less noisy and we can reach better results for high SNR. This indicates that the IQ mismatch coefficients estimation has been improved by using an averaging value from several pairs of pilots.

Figures 4.20 and 4.21 show 64-QAM constellations demodulated without and with IQ mismatch compensation, respectively. In addition we plotted on the right-hand side of both figures the PSDs of the WiMAX signal and the ICI noise due to the RF mixer IQ mismatch. The receiver gain and phase imbalances have been set to 1 dB (12.2%) and $3°$, as in the two previous simulations, and an AWGN channel limits the maximum SNR around 40 dB; that is, minimum EVM around -40 dB.

Without the compensation, the constellation is noisy with an EVM around -23 dB fixed by the receiver IQ mismatch. We can see in the spectral plot the ICI noise generated by the mixer IQ mismatch within the signal band, which is much higher than the AWGN. With the compensation one can see that the constellation points become cleaner with an EVM improvement of ∼15 dB even if the ICI noise due to the RF analog IQ mismatch is still present, as shown in the spectral plot.

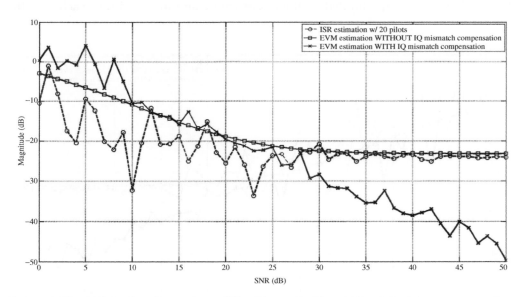

Figure 4.19 EVM estimation versus SNR with and without IQ mismatch compensation using 20 pilots

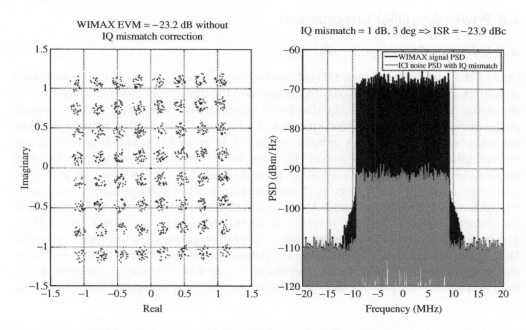

Figure 4.20 WiMAX 64-QAM constellation and PSD without IQ mismatch compensation

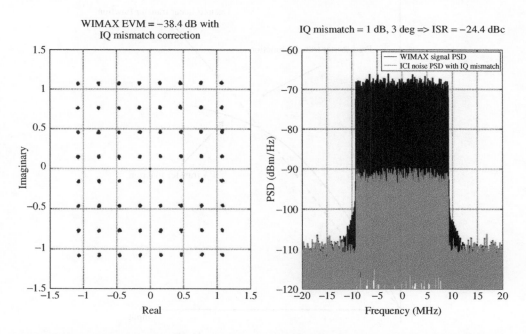

Figure 4.21 WiMAX 64-QAM constellation and PSD with IQ mismatch compensation

4.5 Power Amplifier Linearization

4.5.1 Digital Predistortion Principle

One of the major issues with OFDM modulation is the high PAPR of the temporal signal, which dictates the linearity requirement of the transceiver chain. In transmission the signal distortion is especially due to PA nonlinearity, which degrades the system performance and also generates out-of-band emissions. The simplest solution to limit the transmission distortion is to back-off the PA in order to operate in a wide linear region at the detriment of the power efficiency. However, in modern integrated CMOS transceivers, PAs are sometimes also embedded on-chip for cost reasons and then cannot take advantage of large back-off due to a limited voltage supply. So a new design tendency is to allow the PA to operate in its nonlinear region and to compensate the distortion using digital predistortion (Cavers, 1990; Ding *et al.*, 2004; Li *et al.*, 2009). The basic principle of the compensation is to predistort the signal in DBB before the DAC, that is, ahead of the PA, with a function that is the inverse of the nonlinear transfer function of the PA, as represented in Figure 4.22. Whereas PA compresses at high amplitudes, the predistorter will compensate the compression with an expansion.

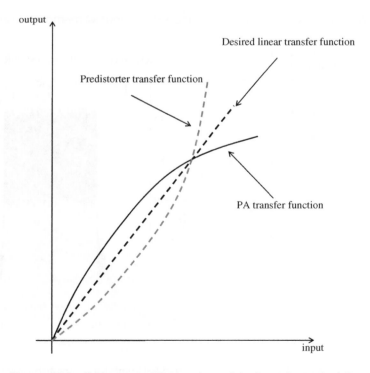

Figure 4.22 Power amplifier and predistortion characteristics

Furthermore, because the PA transfer function can vary in time, for example, due to temperature changes or output impedance variation, adaptive digital predistortion can be introduced to track these variations. In that case the predistorter transfer function is computed in DBB using the input and output baseband signals of the PA fed back, as depicted in Figure 4.23. This technique is known as an indirect learning architecture-based predistorter.

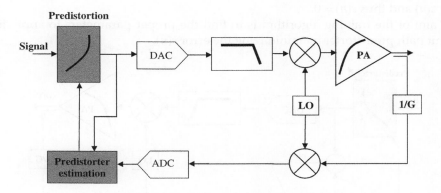

Figure 4.23 Power amplifier adaptive predistortion principle

4.5.2 Memory Polynomial Predistortion

In modern wideband OFDM communication systems, early predistortion techniques using memoryless models became ineffective as PA memory effects were no longer negligible and had to be taken into account in the compensation scheme. Consequently, to efficiently linearize a PA with memory, the predistorter must integrate the memory behavior into its model. One solution to model the PA memory is to use polynomials with sample history (Ding *et al.*, 2002, 2004; Li *et al.*, 2009):

$$y(n) = \sum_{k=1}^{K} \sum_{q=0}^{Q} a_{kq} x(n-q) \left| x(n-q) \right|^{k-1} \tag{4.55}$$

where $x(n)$ and $y(n)$ represent the input and the output of the PA, respectively, Q is the memory length, K is the highest order of the polynomial, and a_{kq} are the coefficients of the polynomial. The predistortion performance strongly depends on the estimation accuracy of the polynomial coefficients calculated in the feedback path.

It is interesting to note that if we assign $Q=0$ in Equation 4.55, that is, no memory, we obtain

$$y(n) = \sum_{k=1}^{K} a_k x(n) \left| x(n) \right|^{k-1} \tag{4.56}$$

which is the conventional memoryless polynomial used to model a PA without memory.

In order to compute the polynomial coefficients a_{kq} of Equation 4.55, an indirect learning architecture is generally used to train the predistorter (Ding *et al.*, 2002, 2004; Li *et al.*, 2009). Figure 4.24 shows the principle of PA predistortion with an indirect learning architecture. The output of the PA is attenuated and fed back to the "predistorter training" block which computes the predistorter coefficients by minimizing the error between the predistorded signal and its estimated value from the training block. For an ideal predistortion, we should have $y(n) = Gx(n)$, giving $\hat{z}(n) = z(n)$ and thus $e(n) = 0$.

The aim of the training algorithm is to find the proper parameters to apply in the transmit path predistorter by minimizing the error $e(n)$.

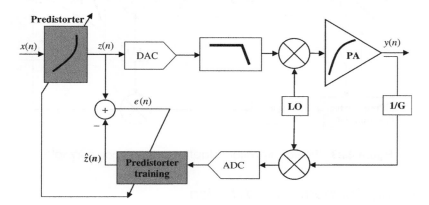

Figure 4.24   Power amplifier predistortion with indirect learning architecture

4.5.3 Polynomial Coefficients Computation

Because the memory polynomial which generates $y(n)$ as defined by Equation 4.55 is linear in coefficient a_{kq}, the latter can be estimated using a least-squares algorithm (Ding *et al.*, 2004).

The predistorter training algorithm has to compute the coefficients from the PA output data (Figure 4.24), and consequently Equation 4.55 has to be rewritten:

$$z(n) = \sum_{k=1}^{K}\sum_{q=0}^{Q} a_{kq}y_{kq}(n) \tag{4.57}$$

with

$$y_{kq}(n) = \frac{y(n-q)}{G}\left|\frac{y(n-q)}{G}\right|^{k-1} \tag{4.58}$$

In order to simplify the notation, it is preferable to use the matrix form to express Equation 4.57:

$$\mathbf{z} = \mathbf{Ya} \tag{4.59}$$

where for N samples we have

$$
\begin{aligned}
\mathbf{z} &= [z(0), \cdots, z(N-1)]^\mathrm{T} \\
\mathbf{Y} &= [\mathbf{Y}_0, \cdots, \mathbf{Y}_Q] \\
\mathbf{Y}_q &= [\mathbf{y}_{1q}, \cdots, \mathbf{y}_{Kq}] \\
\mathbf{y}_{kq} &= [y_{kq}(0), \cdots, y_{kq}(N-1)]^\mathrm{T}
\end{aligned}
\tag{4.60}
$$

and the memory polynomial coefficient vector

$$
\mathbf{a} = [a_{10}, \cdots, a_{K0}, \cdots, a_{1Q}, \cdots, a_{KQ}]^\mathrm{T}
\tag{4.61}
$$

In steady state the least-squares solution for Equation 4.59 gives the estimate of the memory polynomial coefficients (Ding *et al.*, 2004; Li *et al.*, 2009):

$$
\hat{\mathbf{a}} = (\mathbf{Y}^H \mathbf{Y})^{-1} \mathbf{Y}^H \mathbf{z}
\tag{4.62}
$$

where $(.)^H$ denotes the Hermitian transpose. The dimensions of the matrix $\mathbf{Y}$ and vector $\hat{\mathbf{a}}$ are directly related to the number of samples N, the polynomial order K, and memory length Q: $[N \times K(Q+1)]$ for $\mathbf{Y}$ and $[K(Q+1) \times 1]$ for $\hat{\mathbf{a}}$.

The accuracy of the polynomial coefficients estimation especially depends on the inverse of matrix $\mathbf{Y}^H \mathbf{Y}$. The condition number of the matrix $\mathbf{Y}^H \mathbf{Y}$ can be used in order to get an indication of the accuracy of the computed solution for the linear system defined in Equation 4.59:

$$
\mathrm{cond}\,(\mathbf{Y}^H \mathbf{Y}) = \frac{\lambda_\mathrm{max}}{\lambda_\mathrm{min}}
\tag{4.63}
$$

where λ_max and λ_min are the maximum and the minimum eigenvalues, respectively, of the matrix $\mathbf{Y}^H \mathbf{Y}$. If the condition number of the matrix is close to one, it is said that the matrix is well conditioned and its inverse can be computed with good accuracy. On the other hand, if the condition number is large then the matrix is poorly invertible and the solution to this linear system may be affected by large numerical errors.

4.5.4 Simulation Results

As for IQ mismatch estimation and compensation algorithm studied in Section 4.4, we simulated the memory polynomial predistortion using the same mobile WiMAX simulation chain described in Chapter 3. We added the predistorter in the transmission path with indirect learning architecture as depicted in Figure 4.24, and the predistorter coefficients were computed using Equation 4.62 for a fifth-order polynomial with a memory length of 3. In practice, the order and the memory length of the polynomial are specific to PA and are determined with measurements.

The simulation consisted of varying the PA output power from 0 to 25 dBm and measuring the EVM with and without digital predistortion. The simulation results

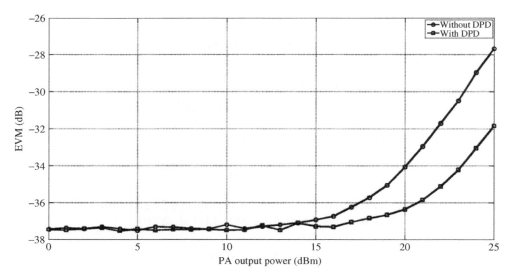

Figure 4.25 Transmitter EVM as a function of the PA output power with and without digital predistortion

are plotted in Figure 4.25, in which we can clearly distinguish the performance improvement at high output powers, higher than 15 dBm, with digital predistortion. For low levels, the EVM is limited to around −37 dB due to the transmission path I and Q mismatch (0.1 dB, 1°) and LO phase noise (PLL bandwidth, 100 kHz; in-band noise, −95 dBc/Hz). Above 24 dBm output power the EVM is enhanced by ∼3 dB, which is significant in terms of power consumption or link budget. Indeed, the performance gain provided by the digital predistortion can be used to obtain better EVM for same output power, or to reduce the transmitter power consumption by improving the PA efficiency, achieving the same EVM at the same output power while using less current.

Figures 4.26 and 4.27 show the transmitter complex constellation and the WiMAX signal power spectral density for 25 dBm output power without and with digital predistortion, respectively. Without digital predistortion (Figure 4.26), the complex constellation exhibits an EVM of −27.8 dB and we noticeably see on the right-hand side of the figure the spectral regrowth due to the PA nonlinearity affecting the adjacent channel selectivity. In Figure 4.27 we can observe the effect and improvement provided by the digital predistortion, as the EVM is improved by 3.9 dB (−31.7 dB) and the out-of-band emission due to spectral regrowth has been significantly reduced. This illustrates that the PA linearity has been enhanced as expected.

Finally, Figure 4.28 shows the predistorter output signal constellation and its power spectral density when the digital predistortion is activated. In effect the predistorted signal is the ideal complex baseband signal distorted with the inverse of the PA nonlinear response. Consequently, the PA input signal exhibits poor performance compared with its output, as characterized by a large spectral regrowth and a lower

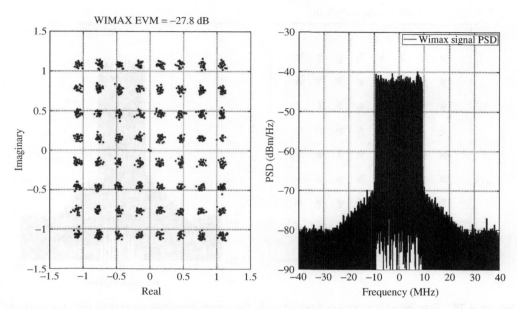

Figure 4.26 Transmitter constellation and PSD without digital predistortion for output power of 25 dBm

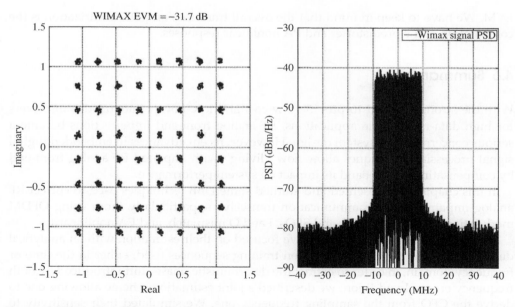

Figure 4.27 Transmitter constellation and PSD with digital predistortion for output power of 25 dBm

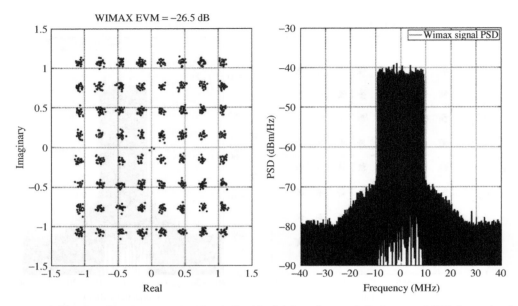

Figure 4.28 Predistorter signal output; that is, PA input: constellation and PSD for output power of 25 dBm

EVM. We have to keep in mind that the overall transmission path linearization is the combination of the predistorter and PA nonlinear responses.

4.6 Summary

With the strong demand to provide low-cost and low-power integrated transceivers for high data-rate mobile applications, RF analog front-end imperfections became a serious concern during system design. However, recent advances in embedded digital signal processing techniques allow now "living" with impaired RF analog front-end by compensating in baseband its impact on system performance.

In this chapter we presented and studied some DBB techniques used to correct RF analog impairments in communication transceivers, particularly those using OFDM modulation, which are carrier and SFOs, I and Q mismatch, and PA nonlinearity.

For carrier and frequency offsets, we focused on their estimation with an analytical description of the algorithms based on training sequences used, either in the time or frequency domain, in order to extract the deterministic phase shift introduced by both frequency errors. In addition, we described a joint estimation scheme allowing one to derive the CFO from the sampling frequency one. We simulated their sensitivity to AWGN and we saw that the estimation of SFO is much more sensitive to noise than that of the carrier frequency.

For IQ mismatch we presented an algorithm which estimates the ISR coefficients in the frequency domain, that is, after the receiver FFT, using at least two known subcarriers (pilots). For a better accuracy, several pairs of subcarriers can be used in order to obtain an average value of the estimate. Afterwards, these coefficients are applied across all the subcarriers so as to compensate the IQ mismatch effect. Simulation results in AWGN confirmed the theory and showed the validity of this technique with a reduction of the ICI noise contribution due to IQ mismatch resulting in an EVM improvement.

Finally, we introduced a DBB predistortion technique for compensating PA non-linearity. The proposed algorithm is based on memory polynomials, and an indirect learning architecture is utilized to compute the predistorter polynomial coefficients to apply to the transmit baseband signal. Simulation results showed an enhancement of the EVM and a reduction of the out-band-emissions for high output power. This improvement can be used to obtain better performance with same PA efficiency or equivalent performance, but with higher PA efficiency by reducing the current consumption.

The research on techniques and algorithms for mitigating different RF analog impairments in OFDM systems is very active and several books and papers deal with this new trend, including Horlin and Bourdoux (2008), Schenk (2008), and Kiayani *et al.* (2012), and references therein.

References

Cavers, J.K. (1990) Amplifier linearization using a digital predistorter with fast adaptation and low memory requirements. *IEEE Transactions on Vehicular Technology*, **39**, 374–382.

Ding, L., Zhou, G.T., Morgan, D.R. *et al.* (2002) Memory polynomial predistorter based on the indirect learning architecture. Proceedings of GLOBECOM, vol. 1, pp. 967–971.

Ding, L., Zhou, G.T., Morgan, D.R. *et al.* (2004) A robust predistorter constructed using memory polynomials. *IEEE Transactions on Communications*, **52**, 159–165.

Fettweis, G., Löhning, M., Petrovic, D. *et al.* (2007) A new paradigm. *International Journal of Wireless Information Networks*, **14** (2), 133–148.

Horlin, F. and Bourdoux, A. (2008) *Digital Compensation for Analog Front-Ends: A New Approach to Wireless Transceiver Design*, John Wiley & Sons, Inc.

Kiayani, A., Anttila, L., Zou, Y., and Valkama, M. (2012) Advanced receiver design for mitigating multiple RF impairments in OFDM systems: algorithms and RF measurements. *Journal of Electrical and Computer Engineering*, **2012**, 16 pp, Article ID 730537.

Lei, W., Cheng, W., and Sun, L. (2003) Improved joint carrier and sampling frequency offset estimation scheme for OFDM systems. *IEEE Global Telecommunications Conference*, **4**, 2315–2319.

Li, B., Ge, J., and Ai, B. (2009) Robust power amplifier predistorter by using memory polynomial. *Journal of Systems Engineering and Electronics*, **20** (4), 700–705.

Moose, P.H. (1994) A technique for orthogonal frequency division multiplexing frequency offset correction. *IEEE Transactions on Communications*, **42** (7) 2908–2914.

Schenk, T. (2008) *RF Imperfections in High-Rate Wireless Systems: Impact and Digital Compensation*, Springer.

Schmidl, T. and Cox, D. (1997) Robust frequency and timing synchronization for OFDM. *IEEE Transactions on Communications*, **45** (12), 1613–1621.

Sliskovic, M. (2001) Carrier and sampling frequency offset estimation and correction in multicarrier systems. IEEE Global Telecommunications Conference, vol. 1, pp. 285–289.

Tubbax, J., Come, B., Van der Perre, L. *et al.* (2003) IQ imbalance compensation for OFDM. IEEE International Conference on Communications, vol. 5, pp. 3403–3407.

Tubbax, J., Come, B., Van der Perre, L. *et al.* (2004) Compensation of transmitter IQ imbalance for OFDM systems, IEEE Proceedings of International Conference on Acoustics, Speech and Signal Processing.

Won, K.H., Han, J.S., and Choi, H-J. (2010) Sampling frequency offset estimation methods for DVB-T/H systems. *Journal of Networks*, **5** (3), 313–320.

Index

---

*RF Analog Impairments Modeling for Communication Systems Simulation:
Application to OFDM-based Transceivers*, First Edition. Lydi Smaini.
© 2012 John Wiley & Sons, Ltd. Published 2012 by John Wiley & Sons, Ltd.

Printed and bound by CPI Group (UK) Ltd, Croydon, CR0 4YY

16/04/2025

14658381-0002